MATERIALS SCIENCE AND TECHNOLOGIES

COMPUTATIONAL MATERIALS AND BIOLOGICAL SCIENCES

MATERIALS SCIENCE AND TECHNOLOGIES

Additional books in this series can be found on Nova's website under the Series tab.

Additional e-books in this series can be found on Nova's website under the e-book tab.

MATERIALS SCIENCE AND TECHNOLOGIES

COMPUTATIONAL MATERIALS AND BIOLOGICAL SCIENCES

KHOLMIRZO T. KHOLMURODOV
EDITOR

nova publishers
New York

NOTICE TO THE READER

Additional color graphics may be available in the e-book version of this book.

Library of Congress Cataloging-in-Publication Data

Computational materials and biological sciences / editor, Kholmirzo T. Kholmurodov (Leading Scientist, Frank Laboratory of Neutron Physics, Joint Institute of Nuclear Research, Dubna, Moscow Region, Russia).
pages cm. -- (Materials science and technologies)
Includes index.
ISBN 978-1-63482-541-2 (hardcover)
1. Molecular biology--Computer simulation. I. Kholmurodov, Kholmirzo.
QH506.C6418 2015
572.80285--dc23

2015010476

Published by Nova Science Publishers, Inc. † New York

CONTENTS

Preface

In this book, the original papers have been collected that demonstrate the efficient use of the computer molecular dynamics (MD) simulation methods for the studying of nanoscale phenomena in the materials and life sciences. This volume contains the Proceedings of the MSSMBS-2014 and DSCMBS-2014 International Workshops (7th Japan–Russia Workshop "Molecular Simulation Studies in Materials and Biological Sciences" and 1st Japan–Russia–Tajikistan Symposium "Dushanbe Symposium on Computational Materials and Biological Sciences"), which have been organized by the Joint Institute for Nuclear Research (JINR, Dubna), the Institute of Bioorganic Chemistry of the Russian Academy of Sciences (IBCh, RAS, Moscow) and S.U. Umarov Physical-Technical Institute of the Academy of Sciences of the Republic of Tajikistan (PhTI, AS, RT) during the September 21–28, 2014. The MSSMBS'14 & DSCMBS'14 workshops got started in IBCh, Moscow, Russia, and finished in PhTI, Dushanbe, Tajikistan.

One has to mention that the Japan-Russia International Workshop MSSMBS with active participation of European, USA, Canadian, etc., scientists and research groups has already been held five times during 2004-2012 at JINR (the 5th MSSMBS workshop was held on the base of JINR of Dubna on 9-12 September 2012). Also during 10-12 September 2013 in Japan the 1st International Symposium on Computational Materials and Biological Sciences (ISCMBS) was held in Waseda University of Tokyo, that make the bilateral collaboration more international. Thus, the border of the scientific cooperative meeting on the world scale has greatly been expanded, with a hope that the DSCMBS Symposium in Dushanbe will provide a good chance for the opening of the trilateral communication in science between Tajikistan, Japan and Russia.

So far, in September 2014 the delegation of world leading scientists from Russia, Japan and other countries make a first visit to Tajikistan capital and held a new international symposium DSCMBS-2014, hosted by S.U. Umarov Physical-Technical Institute of the Academy of Sciences of the Republic of Tajikistan. The Organizing committee was as: Kholmirzo T. Kholmurodov - Cochairman from JINR, Dubna, Moscow region, Russia; Yuko Omamoto - Cochairman from Nagoya University, Japan; Roman G. Efemov - Cochairman from IBCh, Russian Academy of Sciences, Moscow, Russia; Farhod K. Rakhimov - Cochairman from Academy of Sciences of the Republic of Tajikistan, President of Tajik Academy of Sciences; Khikmat Kh. Muminov - Cochairman from S.U. Umarov Physical-Technical Institute, Director of this institute.

The DSCMBS-2014 symposium has appeared to provide an ideal opportunity to discuss the latest development and exchange technical ideas in the field of computational materials and biological sciences. Especially, as organizers were planning, the DSCMBS-2014 would promote to establish the very close cooperation between young researchers that help to strengthen the future scientific cooperation between Tajikistan, Russia and Japan. In the framework of the DSCMBS the presentation talks and lectures were given not only by leading Russian and Japanese scientists, but also by young researchers from Russia, Japan, European and other countries, covering the following topics: the development of high performance computers and new theoretical methods; computational methods in modern materials and biological sciences, and so on. The contributions of Tajikistani scientists were greatly welcome thereby making their bridges with Japanese and Russian colleagues and to promote a new branch of world science in the own country.

In particular, the research topics that covered at DSCMBS-2014 were as follows:

*Computer molecular simulation methods and approaches

*Molecular dynamics (MD) and Monte-Carlo (MC) techniques

*Modeling of biological molecules (RNA, DNA, proteins, enzymes, mutation transitions, etc.)

*Physical and biochemical systems (gasses, crystals, liquids, polymers, biotribology, etc.)

*Material fabrication and design (ion doped structures, high pressure clathrates, carbon nanotubes, etc.)

*Drug design in medicine (docking of one molecule to another, inhibitory activities of enzymes, etc.)

*Computational and computing physics, chemistry, biology and medicine

*GPU accelerated molecular dynamics & related techniques

The representatives of the research institute and universities that participated the MSSMBS-2014 & DSCMBS-2014 workshops included: Keio, Waseda, Nagoya and Kyoto Universities, Nagahama Institute of Bio-Science and Technology (Japan), Lomonosov Moscow State University, Joint Institute for High Temperatures (RAS, Moscow), Puschino Scientific Center (RAS, Puschino), Institute of Bioorganic Chemistry (RAS, Moscow), National Academy of Sciences (Armenia), University Paris-Diderot (France), so on.

Such high level Japan-Russia-Tajikistan conferences with active contributions of scientists from Europe, USA and other countries are focused on the computational simulation, mainly on the molecular dynamics simulation, thereby serving as an example of rapid informational and technological exchanges in our rather fast changing world of creation of new materials, new drugs and new scientific-educational tools. It is also important to note the cultural and humanitarian exchanges between the delegations of the participating countries on the MSSMBS-2014 and DSCMBS-2014 International Workshops.

Kholmirzo T. Kholmurodov
Organizing Chairman
The MSSMBS'14 & DSCMBS'14 International Workshops
Joint Institute for Nuclear Research
6 Joliot-Curie St., Dubna, 141980, Moscow Region, Russia
E-mail address: mirzo@jinr.ru

Computational Materials and Biological Sciences: An Introduction

During September 21–28, 2014 the Joint Institute for Nuclear Research (JINR, Dubna, Moscow Region, Russia) with the Institute of Bioorganic Chemistry of Russian Academy of Sciences (IBCh, RAS, Moscow, Russia) and S.U. Umarov Physical-Technical Institute of Academy of Sciences of the Republic of Tajikistan (PhTI, AS, RT) and in close cooperation with japanese scientists hosted the 7th Japan–Russia International Workshop MSSMBS-2014 "Molecular Simulation Studies in Material and Biological Sciences" and 1st Japan–Russia–Tajikistan International Symposium DSCMBS-2014 "Dushanbe Symposium on Computational Materials and Biological Sciences" (co-chaired by Profs. Kholmirzo Kholmurodov, Roman Efremov, Yuko Okamoto, Farhod Rakhimov and Khikmat Muminov). The MSSMBS'14 and DSCMBS'14 workshops provided a lot of interesting talks on different aspects of materials science and biological research and outlined the current state and prospects of computer molecular simulation studies; it covered a broad spectrum of problems in modern physics, biochemistry, and nanotechnology.

In the MSSMBS-2014 & DSCMBS-2014 International Workshops the following topics, in particular, were discussed: computer molecular simulation methods and approaches; molecular dynamics (MD) and Monte-Carlo (MC) techniques; modeling of biological molecules (RNA, DNA, proteins, enzymes, mutation transitions, etc.); physical and biochemical systems (gasses, crystals, liquids, polymers, bio-tribology, etc.); material fabrication and design (ion doped structures, high pressure clathrates, carbon nanotubes, etc.); drug design in medicine (docking of one molecule to another, inhibitory activities of enzymes, etc.); computational and computing physics, chemistry, biology and medicine; GPU accelerated molecular dynamics & related techniques.

The MSSMBS'12 & DSCMBS'12 International Workshops were represented by leading computer molecular simulation research groups of Japan (Keio, Waseda, Nagoya and Kyoto Universities, Nagahama Institute of Technology and Science, etc.), Russian Federation (JINR, M.V. Lomonosov Moscow State University, St.Petersburg Nuclear Physics Institute, Institute of Bioorganic Chemistry, the Joint Institute for High Temperatures of the Russian Academy of Sciences, Puschino Scientific Center of the Russian Academy of Sciences), Republic of Armenia (National Academy of Sciences), France (University Paris-Diderot), USA, Europe and other countries.

In the first paper by M. Matsumoto and M. Ide the author developed a simple particle model for a polymer brush system in order to examine tribological behavior of articular cartilage in human joints. The brush model consists of coarse-grained beads connected with harmonic springs, grafted on a smooth plate. The Lennard-Jones type short-ranged interaction is assumed between beads. Random forces and hydrodynamic frictions proportional to the relative velocity are taken into account, where the local velocity of surrounding fluid is determined with the mean local density of beads. A series of molecular dynamics simulations has been carried out with this model to investigate the polymer deformation and the frictional force acting on the plates.

In the the second paper by A. Shvetsov, D. Lebedev and V. Isaev-Ivanov molecular mynamics (MD) simulations were used together with existing X-ray structures to build the structures of RecA proteins. The author built missing loops L1 and L2 of RecA from *D. radiodurans* using ICM homology and loop modeling modules. The structure verification via SANS experiments was performed to build realistic all-atom models of quite large macromolecular multicomponent complexes like it was done for RecA::RecX::ssDNA and RecX::ssDNA complexes (this structure was predicted via MM and MD and then verified by SANS).

In the 3rd paper by R. Eremin, Kh. Kholmurodov, V. Petrenko and M. Avdeev the limiting solutions of oleic acid (OA) in deuterated benzene and decalin were previously studied by molecular dynamics (MD) simulations.

meability, but at the same time liquid-crystalline state, which is vital for every living cell.

In the 10th paper E. Dushanov and Kh. Kholmurodov the behavior of DNA molecule in a repairing molecular system was discussed. As repairing molecular system the author simulated an *Anacystis nidulans* photolyase enzyme that contains two chromophore cofactors: flavin adenine dinucleotide (FAD; flavin) and 8-hydroxy-5-deazaflavin (8-HDF, deazaflavin). This photolyase system was simulated for a photoreaction DNA repairing process via molecular dynamics (MD) method. Observations show that flavin actively participates in a interaction process, thereby involving the conformational changes of the DNA and CPD substrate fragments. Thus, the author observed distinct behaviors between two types of DNA and matching of the average characteristics of healthy DNA with its thymine. This gives an argument to assume that under the influence of active cofactor damaged DNA begins to rotate around its symmetry axis, in order to flip out the CPD.

In the 11th paper by T. Yamamoto, K. Nishimura and T. Yoshioka the analytical method of local environment of dilutely doped magnetic elements in functional materials in an atomic scale by the X-ray absorption near-edge structure (XANES) measurements and the first-principles calculations is demonstrated. Results for the lead-free piezoelectric material, Mn and Fe doped $Bi_4Ti_3O_{12}$, is shown as an application of the above method.

In the 12th paper by M. Khusenov, E. Dushanov and Kh. Kholmurodov the small nucleotide chain (NC) with gold nanoparticles (NPs) inside carbon nanotube (CNT) of an open ended boundary were simulated. Such system represents a great interest in many aspects of today biochemical and nanotechnological research (diagnostic aplications, drug delivery in cell, nanorobotic design and related manipulations). The CNT was described through a quantum-chemistry potential, though the trajectory calculation for the whole NC–NP–CNT system was performed via the classical molecular dynamics (MD) approach. The Lennard-Jones short-ranged interaction is assumed between the NC, NP and CNT. The author have carried out a series of MD simulations with different NC–NP–CNT molecular models to investigate the pequliarities of VdW interactions, bond and structural formations along with the dynamical behavior.

In the 13th paper by S. Kuznetsov, A. Polyansky, P. Volynsky and R. Efremov it is shown that lipid environment promotes dimerization of transmembrane helices. Membrane receptors are among the most important molecules in the cell. Their functionin g determines response of the cell to the external conditions. Also, they provide cellular communication and recognition mechanisms. So, their proper functioning is a vital question of life. There is a lot of data today about these receptors. Many three-dimensional (3D) structures are known, genetic causes of some diseases are determined and many therapeutic agents acting on the membrane proteins are developed. But today with all this knowledge we cannot exactly predict the effects of new mutations in these proteins on their activity and cannot decipher the mechanisms of their action in detail. On the other side, cell membrane consists of phospholipids acting as a matrix for the proteins and other molecules like glycolipids. Of course, exact lipid composition of the membrane affects the proteins behavior, but this phenomenon is reversible. Many papers describe how single-point mutations act on the protein activity, another large area of study includes determination of the influence from the lipid bilayer properties. But little is said about how the protein itself induces local perturbations on the membrane. Taking this into account, the author decided to examine the simplest case of the membrane protein with a help of atomistic computer simulations of molecular dynamics. Using transmembrane (TM) helical peptide representing TM-domain of human glycophorin A in hydrated lipid bilayer, we showed that it can induce local perturbations of lipid properties, and the characteristic size of the pattern is 10-15 Å. So, the peptides can feel each other on large distances (about 20 Å), and this estimation is confirmed by calculation of the association free energy. The author determined that the interaction interface for dimer formation also serves as a binding site for lipids.

In the 14th paper by I. Kovalenko, A. Abaturova, A. Diakonova, D. Ustinin, S. Khruschev, G. Riznichenko and A. Rubin the author develop a method of multiparticle Brownian dynamics and apply it to computer simulation of interactions of electron transport proteins. This method makes it possible to study the effects of various factors on protein-protein complex formation. Some of them are connected with electrostatic interactions (ionic strength and pH of the solution, charge-changing point protein mutations). Other factors are molecule concentration, viscosity of the solution, size and shape of the reaction volume. The author show how Brownian dynamics takes into account these factors by the example of interaction of photosynthetic electron transport proteins plastocyanin and cytochrome f.

In the 15th paper by N. Orekhov and V. Stegailov the experimental data on graphite melting temperature remain poorly determined despite the long history of investigations. The experimental results of several works cover the wide span from 3800 to 5000 K that is an essentially larger uncertainty than the errors of individual experiments. In this work the author deploy the molecular dynamics (MD) method to study the kinetics of graphite melting, concerning the aspects of defect formation and single graphene layer melting. The MD calculations show an unexpectedly weak kinetics of the melting front propagation in graphite that is several orders slower than that in metals. We demonstrate that at sufficiently high heating rates (higher than $10^5 - 10^6$ K/s) the temperatures 500-1000 K above the graphite melting temperature can be reached before the crystal decay. It allows the author to explain long-standing problem of the discrepancy in the experimental data making a hypothesis that there is a strong dependence between experimentally measured graphite melting temperatures and corresponding rates of heating.

In the 16th paper by G. Smirnov and V. Stegailov the author performed coexistence simulations of methane hydrates for pressures up to 5000 bar for different water models. The author calculate the kinetic stability boundary of the superheated metastable sI structure and analyze the effects of the heating rate, system size and cage occupancy. Also molecular dynamics simulation of several possible structures for the new hydrogen hydrate clathrate reported. The author show the strength of molecular simulation as a supplement tool for the analysis of experimental data.

In: Computational Materials ...
Editor: K. T. Kholmurodov, pp. 1-10 ISBN: 978-1-63482-541-2

Chapter 1

LUBRICATION MECHANISM IN JOINTS – ROLE OF BIOPOLYMERS

Mitsuhiro Matsumoto[1,2,*] ***and Mitsuho Ide***[1,3]
[1]Department of Mechanical Engineering and Science,
Graduate School of Engineering,
Kyoto University, Japan
[2]Advanced Research Institute of Fluid Science and Engineering,
Kyoto University, Kyoto, Japan
[3]Toyota Motor Corporation, Toyota, Japan

Abstract

We have developed a simple particle model for a polymer brush system in order to examine tribological behavior of articular cartilage in human joints. The brush model consists of coarse-grained beads connected with harmonic springs, grafted on a smooth plate. The Lennard-Jones type short-ranged interaction is assumed between beads. Random forces and hydrodynamic frictions proportional to the relative velocity are taken into account, where the local velocity of surrounding fluid is determined with the mean local density of beads. We have carried out a series of molecular dynamics simulations with this model to investigate the polymer deformation and the frictional force acting on the plates.

Keywords: Joint Lubrication, Biotribology, Polymer Brush, Coarse Grained Model, MD Simulation, BD Simulation

*E-mail address: matsumoto@kues.kyoto-u.ac.jp

1. Introduction

Animal joints have been a research target from various aspects, such as fundamental tribology (e.g., the mechanism of ultra-low friction under high load), soft matter physics (e.g., dynamic deformation of gel-like materials), and applied biomechanics (e.g., developing better joint replacements) [1, 2]. Essential components of joints are (1) articular cartilage covering each tip of bones, (2) lubricating liquid (i.e., synovial fluid), and (3) joint capsules to confine the liquid. The most relevant one, cartilage, is a very complex tissue, which is a composite of collagen fibers, proteoglycan, and hyaluronic acid. The structure of joints and the mechanism of their lubrication have been widely investigated, but their detail is not well understood yet.

Among the components of cartilage, we focus on the role of proteoglycans, which are glycosylated protein with large negative charge. They are normally combined to the collagen network by hyaluronic acid molecules, and a part of the aggregates exist on the surface of cartilage, forming a polymer brush. The lubrication mechanism of such polymer brush systems has been discussed recently [3, 4]. To investigate the dynamic behavior of the polymer brush on nano-scale level, we developed a coarse-grained polymer brush model and carried out a series of numerical simulations [5].

2. Model and Method

The developed model is schematically shown in Fig. 1; two plates (walls) are slid each other, on which a given number of model polymers are grafted. Periodic boundary conditions are imposed for horizontal directions (x and y).

2.1. Polymer Model

To save the computational time, we adopt a Brownian Dynamic (BD) simulation with a coarse-grained chain model, similar to the Grest-Kremer one [6]. Each polymer, corresponding to a proteoglycan molecule, consists of N beads of point mass m connected with a spring. Four types of forces are exerted on each bead; (1) harmonic spring force, (2) short-ranged bead-bead interaction through a Lennard-Jones (LJ) 12-6 pair potential, (3) bead-wall interaction via a soft-repulsive potential, and (4) hydrodynamic force as a Langevin-type,

$$\vec{F}_{\text{liq}} = -\gamma\left(\vec{v} - \vec{v}_0\right) + \vec{F}_{\text{r}} \tag{2.1}$$

Table 1. Relevant model parameters

Parameter	Symbol	Value	
Number of beads per polymer	N	50	
Equilibrium length of spring	l_0	1.0×10^{-8}	m
Spring constant	k	1.0×10^{-2}	N/m
LJ size parameter of beads	σ	0.89×10^{-8}	m
LJ energy parameter of beads	ϵ	4.0×10^{-22}	J
bead mass	m	2.5×10^{-11}	kg
fluid viscosity	η	0.5	Pa·s
bead friction coefficient	$\gamma \equiv 3\pi\sigma\eta$	4.2×10^{-8}	Ns/m
time step	Δt	1.0×10^{-5}	s

where $\vec{v}$ and $\vec{v}_0$ are the velocity of the bead and the surrounding fluid, respectively. We assume a fluctuation-dissipation theorem between the friction coefficient γ and the amplitude of the Gaussian random force $\vec{F}_r$. Relevant parameters in the model are given in Table 1; detailed descriptions are given in Ref. [5].

Although proteoglycan has a large number of negative charges, they are almost completely screened by counter ions (such as Na^+ and K^+) under normal conditions; the Debye's screening length is an order of one nanometer, much less than the bead diameter. Thus the electrostatic interaction among beads is omitted in this model.

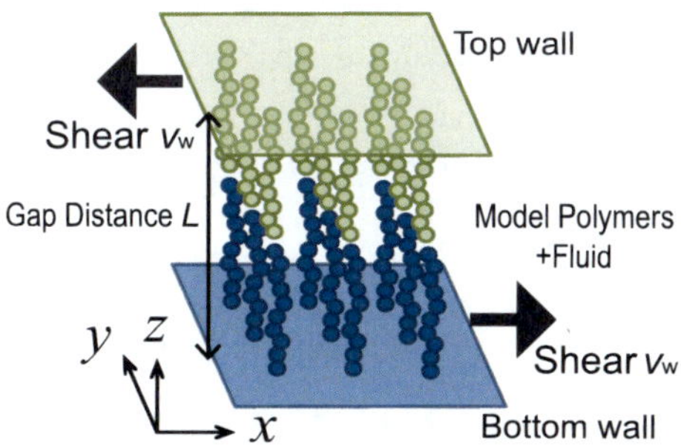

Figure 1. Schematic view of the simulation system.

2.2. Polymer–Fluid Interaction

The local fluid velocity $\vec{v}_0$ in Eq. (2.1) should depend on the simulation conditions. To avoid a time-consuming evaluation of $\vec{v}_0$ based on precise hydrodynamic models, we assume that $\vec{v}_0$ is simply determined by the local bead density as

$$\vec{v}_0 = \vec{v}_C \cdot [1 - \min(\alpha |f_b - f_t|, 1)] + \vec{v}_w \cdot \mathrm{sgn}(f_b - f_t) \cdot \min(\alpha |f_b - f_t|, 1) \quad (2.2)$$

where $\vec{v}_C$ and $\vec{v}_w$ are the Couette flow velocity and the bottom wall velocity, respectively; f_b (f_t) is the packing fraction of beads consisting of grafted polymers on the bottom (top) wall. The numerical factor α corresponds to the effective volume of a bead in reduced unit, and is chosen to be $\alpha = 8.0$ rather arbitrarily. The equation (2.2) implies that each bead of polymers grafted on the bottom (top) wall moves on average with the same speed as the bottom (top) wall, accompanied with the surrounding fluid.

With typical values for the simulation, such as the wall speed $v_w \sim 10\,\mu$m/s and the gap distance $L \sim 500$ nm, the Reynolds number $\mathrm{Re} \equiv \dfrac{\rho v_w L}{\eta}$ (ρ: liquid density, η: liquid viscosity) is an order of 10^{-8}, justifying the implicit assumption of laminar flow in the gap.

2.3. Simulation

With this coarse-grained model, we investigated two systems, a single polymer system and a polymer brush one. All simulations are done at room temperature, $T = 300$ K. Starting from an initial linear configuration of polymers, we thermally equilibrated the system typically for 5×10^6 steps (50 s) before the main calculation.

3. Results and Discussion

3.1. Single Grafted Polymer in Shear Flow

To investigate the properties of the developed coarse-grained model, we first studied deformation of a single polymer under steady shear. The gap distance L is fixed as 500 nm, which equals the length of a fully stretched chain.

Examples of the results are given in Fig. 2, where the height (maximum and mean), the slant angle, and the end-to-end distance are shown. The model

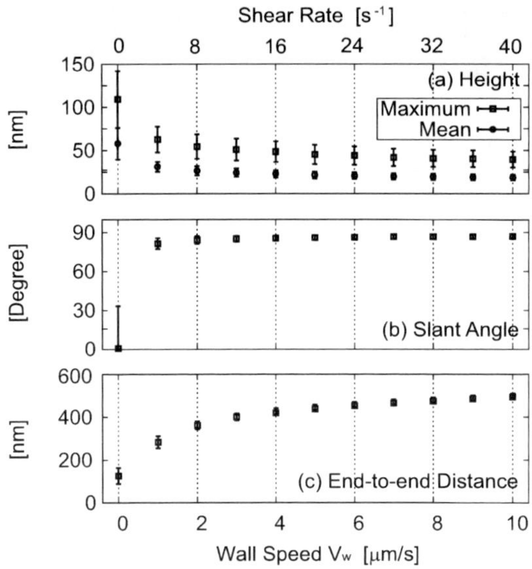

Figure 2. Deformation of a single grafted polymer under steady shear.

polymer is found to be so flexible that even a weak shear of $4\,\mathrm{s}^{-1}$ causes large elongation and strong orientation.

3.2. Polymer Brush in Shear Flow

Two polymer-grafted walls were prepared and slid each other to investigate the tribological properties of a polymer brush system. First, we gradually squeezed the system by reducing the gap distance from $L = 500\,\mathrm{nm}$ to $50\,\mathrm{nm}$ and equilibrated it. Then a constant shear was imposed to see deformation and mechanical response of the brush. Nine polymers are grafted on each wall, and the surface density is varied by changing the surface area, 300×300 and $500 \times 500\,(\mathrm{nm})^2$.

Typical snapshots are shown in Fig. 3, which indicates how the polymer brush deformes as the shear continues. After a sufficiently long run, the system reaches a steady state.

3.2.1. Local Fluid Velocity

As the polymer deformation proceeds, local density of beads changes from the equilibrium one, leading to modification of the fluid velocity $\vec{v}_0$ via Eq. (2.2).

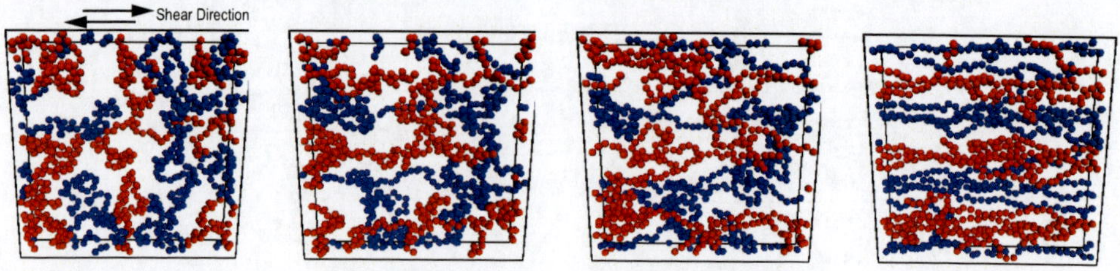

(a) Initial (b) 5,000 steps (c) 10,000 steps (d) 20,000 steps

Figure 3. Snapshots (top view) of the polymer brush simulation; nine polymers in $300 \times 300\,(\mathrm{nm})^2$ with the wall velocity $v_\mathrm{w} = 2.0\,\mu$m/s. Different colors are given to the polymers grafted on the top wall and those on the bottom one.

Examples are shown in Figs. 4 and 5; as the shear increases, the peak of packing fraction shifts to the wall due to the polymer orientation.

3.2.2. Friction on the Wall

Two types of forces are exerted on the walls. One is the force by grafted polymers; another is pressure and shear stress by the fluid.

The polymer force is directly evaluated in the simulation. Since there is no frictional force between the wall and each bead, it is calculated as

$$F_{//} = \sum_{\text{polymer}} f_{//}(0) \tag{3.1}$$

for the tangential component and

$$F_{\perp} = \sum_{\text{polymer}} \left[f_{\perp}(0) - \sum_{i} \frac{\partial \phi_{\mathrm{LJ}}(i)}{\partial z_i} \right] \tag{3.2}$$

for the normal one, where $f(0)$ is the force acting on the grafted end bead and $\phi_{\mathrm{LJ}}(i)$ is the Lennard-Jones interaction potential between the wall and the bead i.

Examples for a transient process are shown in Fig. 6, which is the time evolution of $F_{//}$ and $F_{\perp}$ when the walls start to move at $t = 0$. Since nine polymers of about 500 nm in stretched length are confined (or squeezed) in a narrow gap of $L = 50$ nm, the normal component $F_{\perp}$ takes a fairly large positive

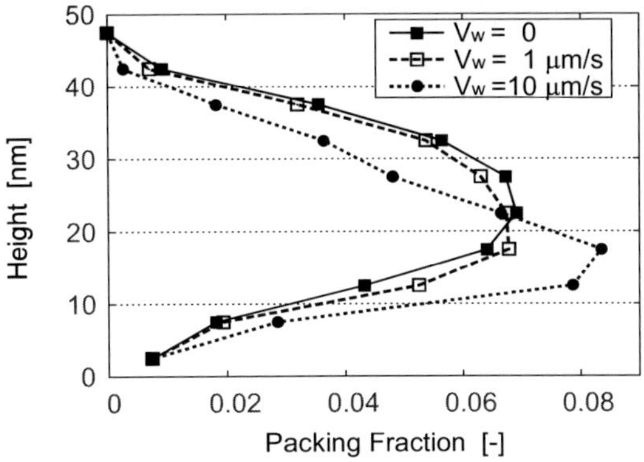

Figure 4. Example of the local packing fraction of polymers grafted on the bottom plate in steady states; nine polymers on $300 \times 300\,(\mathrm{nm})^2$.

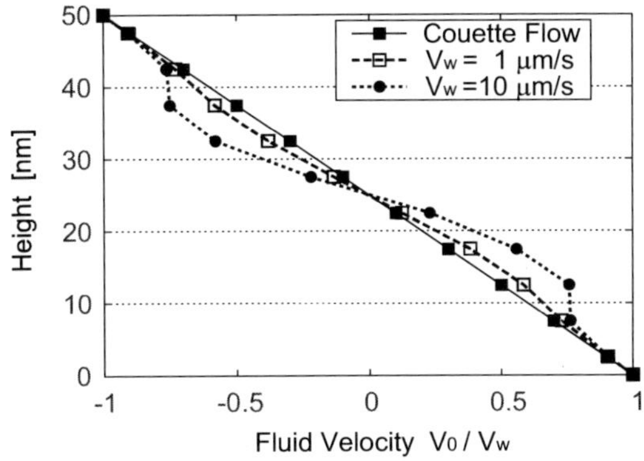

Figure 5. Example of the local fluid velocity in steady states; nine polymers on $300 \times 300\,(\mathrm{nm})^2$.

value even in equilibrium ($v_{\mathrm{w}} = 0$) while the tangential one $F_{//}$ is almost zero. As the polymers stretch with time, however, $F_{//}$ increases and reaches a steady state when the polymers are fully extended.

The averaged force in steady state is shown in Fig. 7. While $F_{//}$, or the frictional force, increases proportionally to the wall speed, $F_{\perp}$ reduces gradually. The graft density seems to little affect $F_{//}$, but small dependence exists in $F_{\perp}$, suggesting that polymer-polymer interaction is not completely negligible.

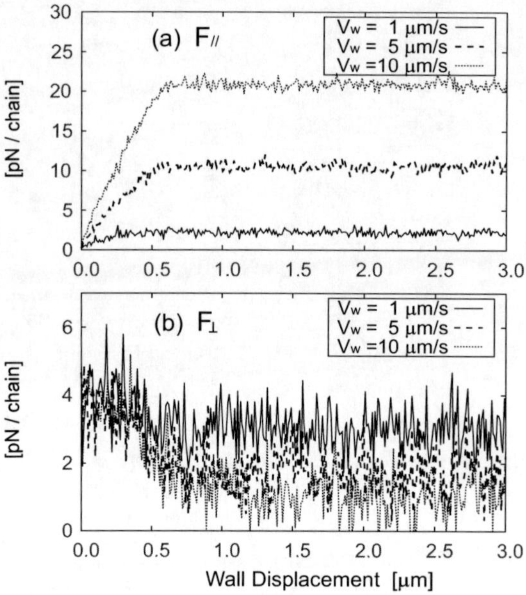

Figure 6. Time evolution of force acting on the walls; nine polymers on $300 \times 300\,(\mathrm{nm})^2$ with $50\,\mathrm{nm}$ gap distance. $F_{//}$: tangential component, $F_{\perp}$: normal component.

3.2.3. Role of Polymer Brush in Lubrication

Based on these simulation results, we can estimate the lubrication force. First, let us discuss the force normal to the cartilage surface, which should be the sum of the polymer force $F_{\perp}$ and the contribution of the synovial fluid pressure; apparently the latter is dominant, and we use a typical value of $1\,\mathrm{MPa}$.

Next, we evaluate the tangential component. The graft density is a relevant parameter, but its value in vivo is not well known; we have $F_{//} \sim 800\,\mathrm{Pa}$ for $v_{\mathrm{w}} = 10\,\mu\mathrm{m/s}$ when the density is rather arbitrarily assumed to be 4×10^9 molecules/cm^2 [or 10 polymers per $500 \times 500\,(\mathrm{nm})^2$] as in the simulation. Although the situation seems different, values of the same order, $F_{//} \sim 500\,\mathrm{Pa}$, is reported with AFM experiments [7]. Synovial fluid also exerts shear stress on the cartilage surface, which is evaluated from the velocity gradient. Since the velocity profile is modified due to the polymer deformation, the shear stress changes from the one in the Couette flow. Under large shears, as shown in Fig. 5

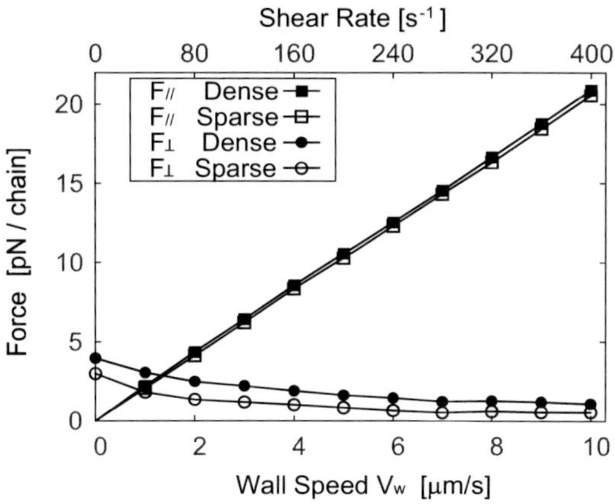

Figure 7. Force components acting on the wall in the steady state; comparison between the dense $[300 \times 300\,(\mathrm{nm})^2]$ case with the sparse one $[500 \times 500\,(\mathrm{nm})^2]$. The typical value of standard deviation is $1 \sim 1.5$ pN/chain.

the velocity gradient near the wall drastically decreases, leading to a large reduction of the shear stress on the wall. In the case of $v_{\mathrm{w}} = 10\,\mu$m/s, the shear stress in the Couette flow of $\sim 200\,\mathrm{Pa}$ reduces by $\frac{1}{5} \sim \frac{1}{2}$ in the existence of polymer brush. As the polymer force and the fluid shear stress have the same order of magnitude, this shear stress reduction is important, which can be one of the main roles of biopolymers on articular cartilages.

With these rough evaluations, we obtain the dynamic frictional coefficient of this model

$$\mu = \frac{\text{tangential force}}{\text{normal force}} \simeq \frac{(800 + 100)\ \mathrm{Pa}}{1\ \mathrm{MPa}} = 9 \times 10^{-4} \tag{3.3}$$

which is comparable to the experimental values, $0.001 \sim 0.01$ [2].

4. Concluding Remarks

We have proposed a coarse-grained polymer brush model to investigate tribology in animal joints. In the simulation, it was suggested that the interaction between synovial fluid and proteoglycan polymers grafted on articular cartilage cause a large modification of local flow from the Couette one, leading to a drastic decrease of shear.

Although this type of bead-spring model is widely used in molecular simulation of polymers, however, it seems too flexible for the aggrecan on the articular cartilage [7]. Some modification based on detailed molecular structure is required. Electrostatic interactions can also have some role in the polymer flexibility. Another important factor is the viscous property of the synovial fluid. Since it is an aqueous solution of polymer electrolytes of large molecular weight, such as hyaluronic acid and lubricin, it shows shear thinning under high shear rates. This non-Newtonian property certainly plays a role in joint lubrication [8]. Further model improvement from these aspects is under way.

References

[1] Van C. Mow and W. M. Lai, "Mechanics of animal joints," *Ann. Rev. Fluid Mech.*, 11 (1976) 247–288.

[2] D. Dowson, "Bio-tribology," *Faraday Discuss.*, 156 (2012) 9–30.

[3] J. Klein, "Polymers in living systems: From biological lubrication to tissue engineering and biomedical devices," *Polymers Adv. Tech.*, 23 (2012) 729–735.

[4] G. W. Greene, A. Olszewska, M. Osterberg, H. Zhu and R. Horn, "A cartilage-inspired lubrication system," *Soft Matter*, 10 (2014) 374–382.

[5] M. Ide and M. Matsumoto, "Tribology of polymer brush: Microscale modelling and simulation," *Molecular Simulation* (2014) DOI: 10.1080/08927022.2014.913790

[6] M. Murat and G. S. Grest, "Structure of a grafted polymer brush: A molecular dynamics simulation," *Macromolecules*, 22 (1989) 4054–4059.

[7] L. Han, D. Dean, C. Ortiz and A. J. Grodzinsky, "Lateral nanomechanics of cartilage aggrecan macromolecules," *Biophys. J.*, 92 (2006) 1384–1398.

[8] A.Y. Hui, W.J. McCarty, K. Masuda, G.S. Firestein and R.L. Sah, "A system biology approach to synovial joint lubrication in health, injury, and disease," *Systems Biology Medicine*, 4 (2012) 15–37.

In: Computational Materials ... ISBN: 978-1-63482-541-2
Editor: K. T. Kholmurodov, pp. 11-23

Chapter 2

Using the Molecular Dynamics Simulation and Neutron Scattering Methods to Study Structural and Functional Properties of Biological Macromolecules

***Alexey Shvetsov*[1,2,*], *Dmitry Lebedev*[1] and *Vladimir Isaev-Ivanov*[1]**
[1]FSBI "Petersburg Nuclear Physics Institute",
NRC "Kurchatov Institute", Gatchina, Russia
[2]Institute of Applied Mathematics and Mechanics,
St. Petersburg State Polytechnical University,
Saint-Petersburg, Russia

Abstract

Molecular Dynamics (MD) simulations were used together with existing X-ray structures to build the structures of RecA proteins[10, 2]. We built missing loops L1 and L2 of RecA from *D. radiodurans* using ICM[11, 12] homology and loop modeling modules. The structure verification via SANS experiments was performed to build realistic all-atom models of quite large macromolecular multicomponent complexes like it was done for RecA::RecX::ssDNA and RecX::ssDNA complexes (this structure was predicted via MM and MD and then verified by SANS).

*E-mail address: alexxy@omrb.pnpi.spb.ru

Keywords: MD Modeling, Veryfication, Full-Atom Structure, Dynamics, RecA and RecX Complexes

1. Introduction

1.1. Small Angle Scattering

The advantage of the methods of small-angle X-ray (SAXS), neutron scattering (SANS) and neutron spin echo (NSE), in the study of biological macromolecular systems is the ability to conduct research in aqueous solution, that is, in conditions close to native. A biological macromolecular structure in time can exist in different conformational states. Therefore, the observable scattering curve, is the result of scattering on the structure averaged over time on the conformational state of the system under study. With this in mind, when solving the direct problem - all-atom molecular modeling study of structures, it is necessary to use modern methods of molecular dynamics (MD) simulations in explicit solvent whose trajectory over time and is a set of structures (frames) reflecting the conformational dynamics of the system under study. In the experiment, SANS and SAXS registers the dependency of the scattering intensity of the value of the scattered vector. This scattering curve using Fourier transform can be transformed to the distance distribution function (DDF) for the scattering object. Modern SANS spectrometers allow to measure the intensity of neutron scattering in a wide range of the scattering vectors, $5 \cdot 10^{-1} \ldots 7 \cdot 10^{-5} \mathring{A}^{-1}$. This range corresponds to the linear dimensions of the scattering object by $10\mathring{A} \ldots 10\mu m$. An advantage of SAXS is its ability to work with a small amount of experimental samples (as low as several μ l). On the other hand, SANS contrast variation technique can be easily accomplished by changing the ratio of D_2O / H_2O, which allows one to separate the contributions of various components of the investigated structures of macromolecular complexes. Since verification of the proposed methods of molecular modeling of structures is carried out with experiments on small-angle scattering, the first step in such verification should be calculation of the scattering curve obtained for the simulated structure. The calculation for SAXS and SANS must be performed by averaging over molecular dynamics trajectory frames. The calculation of the relaxation curves NSE (I(Q,t)) based on the frame MD can be carried out by Fourier transform of the space-time cross-correlation functions.

2. Modeling and Veryfication of Full-Atom Structure and Dynamics of RecA and RecX Complexes

RecA protein [1, 2] plays key role in homology recombination reaction in bacterial cell. It works by forming right handed filament structure on ssDNA[3]. RecA searches homology region between two DNA molecules: ssDNA inside RecA filament and dsDNA outside it, and then it does strand exchange reaction between this two homology regions.

RecX, a small bacterial protein, is known as a negative regulator of the homologous recombinase RecA which is the essential part of the DNA reparation machinery in bacteria. RecX has been shown to bind to RecA filament and inhibit its activity of the homologous strand exchange. An ability of RecX to form a filament by binding to the grove on *RecA::ssDNA* presynaptic complex have been implied in the earlier cryo-EM studies[4].

2.1. Modeling of RecA and RecX Protein Complexes

The X-ray crystallography structure of RecA protein and their complexes have substantial limitations in explaining the functional behaviour of this enzyme on the molecular level. It has been shown that the filaments formed by RecA have strong large-scale conformational flexibility so that their geometry in solution can be quite different from the crystallographic structure [5, 6, 7, 8]. Such flexibility is likely related to the function of the enzyme, particularly with its the interaction with substrate. The structures of the DNA complexes of the enzyme [3] confirms the filament stretching upon DNA binding, but does not address the C-terminal domain conformation, which seems to be a major functional motion in the enzyme [9].

To build starting structures for Molecular Dynamics (MD) simulations we used existing X-ray structures of RecA proteins[10, 2]. We built missing loops L1 and L2 of RecA from *D. radiodurans* using ICM[11, 12] homology and loop modeling modules. Resulting RecA protein structures shown on figure 1. To build model of RecA from *D. radiodurans* presynaptic complex we used existing X-ray structure of RecA from *E. coli* presynaptic complex[3] as homology structure template.

There is no known X-ray structure of *RecA::RecX* exist at this moment, the only known low resolution structure is based on Cryo-EM technique[4]. We

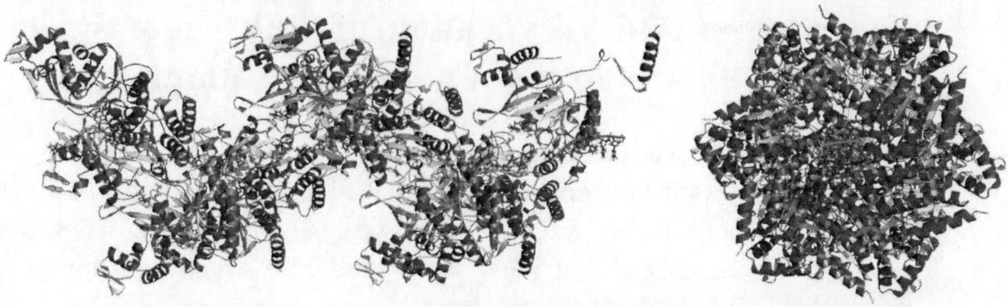

Figure 1. Reconstruction of full-atom model of *RecA::ssDNA* complex.

started from structure of *RecA::ssDNA* complex[3] and structure of RecX[13] protein. Since Cryo-EM data shows that RecX should bind to *RecA::ssDNA* complex filament groove we built a set of starting structures for md simulation using protein-protein docking with different stoichiometric ratios of RecX to RecA.

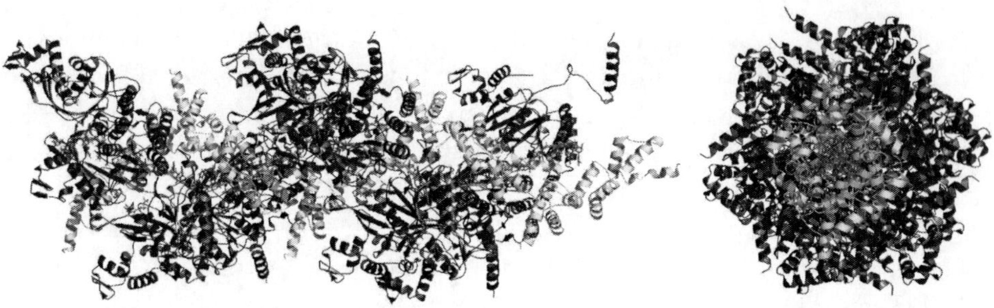

Figure 2. Reconstruction of full-atom model of *RecA::RecX::ssDNA* complex. RecX colored yellow, RecA colored blue..

To model RecX::ssDNA complex we have noticed that RecX can form a chain of head-to-tail dimers that has positively charged surface on one side. We proposed such bend-like RecX filaments can bind to ssDNA in two modes, from one side of ssDNA and from both sides forming sandwich-like structure (figure 3).

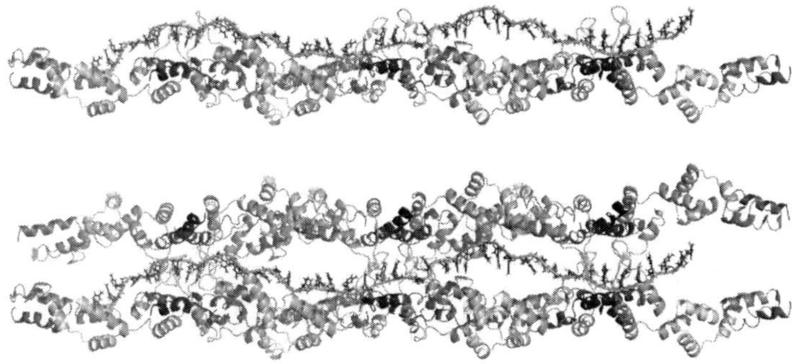

Figure 3. Full-atomic model for RecX::ssDNA chain (top) and sandwich-like filament (bottom).

3. Molecular Dynamics of RecA::RecX Complexes

All MD simulations was performed using GROMACS[14] software package using amber99sb-ildn force field for protein and dna, and tip3p model for explicit water. Simulated systems were around 700k atoms in size including explicit solvent shell. Solvent shell thickness was 1.5nm at minimum. A total of 100 ns were simulated with a time step of 2 fs. Neighbor searching was performed every 50 steps. The S-PME algorithm was used for electrostatic interactions with a cut-off of 1.2 nm. A cut-off of 1.2 was used for Van der Waals interactions. Temperature coupling was done with the Nose-Hoover algorithm. Pressure coupling was done with the Parrinello-Rahman algorithm.

4. SANS Experiment and Model

4.1. Simulation of SANS Spectra from MD

To simulate SANS spectra from MD trajectory we used a version of *g_sans*[9] tool enhanced to support MD trajectory averaging. The methodology difference between static spectra computation and trajectory averaging can be described as following. Since scattering intensity can be described as

$$S(\vec{q}) = \int G(\vec{r}) e^{i\vec{q}\vec{r}} d\vec{r}$$

First step will be getting radial distribution function $G(r)$ for each trajectory frame:

$$G(r) = \sum_{\substack{i,j \\ r=\|\vec{r_i}-\vec{r_j}\|}} b_i b_j$$

where b_i and b_j are scattering length for particles i and j. Then we do averaging of $G(r)$ over all frames of MD trajectory.

This method [15] allows to compute SANS spectra for multimolecular complexes taking into account the conformational changes in solution. While it reproduces the results of the existing packages for static structures, for flexible structures such as RecA filaments it appears to yield a more realistic results than just a static structure approximation (fig 4 left panel).

4.2. SANS Experiments

Small angle neutron scattering experiment was employed to obtain the information on the low-resolution structures of RecX::RecA::ssDNA and RecX::ssDNA complexes in solution, and to test whether molecular model of such an interaction is supported by experiment. Presynaptic RecA filaments were formed on a single-stranded oligonucleotide 104 bases in length and stabilized by a non-hydrolyzable ATP analogue, ATPγS, in accordance with the earlier biochemical studies[16]. SANS spectra were registered on KWS-2 spectrometer at Muenchen FRM-2 reactor.

In the range of scattering vector magnitudes above 0.01 Å^{-1} the experimental spectrum fit very well to the MD simulation for the complex of 33 RecA molecules with 104-long single stranded oligonucleotide (figure 1). RecX interaction with the presynaptic complex did not affect the rod-like shape of RecA polymer, but significantly reduced the diffraction peak near 0.1 Å^{-1} (Figure 4). The effect of RecX is much larger than would be expected from a mere addition of scattering from a RecX filament on single-stranded oligonucleotide and can be fit fairly well to the simulated curve obtained from an MD model for *RecA::RecX::ssDNA* complex at 2:5 binding stoichiometry (figure 2).

SANS spectrum of the complex formed by RecX on ssDNA (figure 4) in the range above 0.02 Å^{-1} was in a good agreement with the model spectrum calculated from MD simulation of the sandwich-like structure and could not be described by a single-sided model.

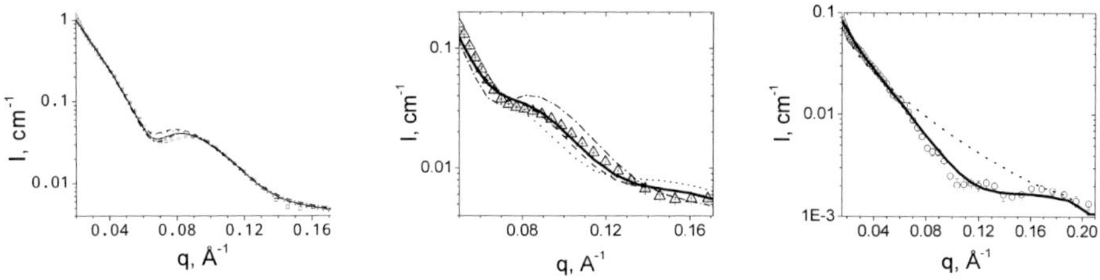

Figure 4. SANS spectrum (left panel) of RecA complex with ssDNA (presynaptic complex). Lines show the model spectra calculated from MD simulation of RecA::ssDNA::ATP complex by GROMACS g_sans[15] tool with (solid) and without (dash-dot line) taking DNA contribution into account, and from the starting (10 ns equilibrated) structure of MD by CRYSON[17] (dashed) and GROMACS utility g_sans (dot line). Data were fitted to the simulated spectra in the range 0.02 - 0.2 $^{-1}$ to adjust for incoherent background and protein concentration. The figure is adapted from [18]. Discrimination between different models for RecA and RecX complexes by SANS. Scattering by RecX complex with RecA presynaptic filament (middle panel) is best fitted to the 2:5 stoichiometry complex (solid line) and differs sharply from the calculated curve for the presynaptic complex model (dash-dot line) and models with different stoichiometry of RecX binding (dashed and dotted lines). Scattering by RecX in the presense of ssDNA (right panel) can be fitted to the sandwich (solid line) but not the one-sided binding (dotted line) model.

5. Spin-Echo Experiment and Model

5.1. Simulation of NSE Spectra from MD

To simulate NSE spectra from MD trajectory we used modified version of *g_sans*[9] tool called *g_nse* enhanced to support MD trajectory averaging and NSE spectra calculation. The normalized spin-echo signal can be calculated as follows:

$$I(\vec{q},\tau) \propto \langle \sum_i^N \sum_j^N \exp(-i\vec{q}(\vec{r_i}(t) - \vec{r_j}(t+\tau))) \rangle$$

where $\langle\rangle$ denotes to averaging over ensemble that equivalent to averaging over t, this equation can be rewritten as following

$$I(q,\tau) = \int P(r,\tau)\frac{\sin(qr)}{qr}dr$$

where

$$G(r,\tau) = \sum_{\substack{i,j \\ r=\|\vec{r_i(t)}-\vec{r_j(t+\tau)}\| \\ \tau=\|t_i-t_j\|}} b_i b_j$$

where b_i, b_j are scattering length on particles i and j

$$P(r,\tau) = \frac{G(r,\tau)}{\int_0^\infty G(r,\tau)dr}$$

As result we get series of $I(q,\tau)$ curves that approximate the averaged NSE relaxation curves at different scattering angles.

5.2. NSE Experiment on RecA Protein from D. Radiodurans

NSE relaxation was measured on IN-15 spectrometer (ILL, Grenoble, France) for 12 mg/ml aqueous solution of RecA protein from D. radiodurans in a D2O buffer containing ca.5% glycerol. Measurements were performed for the scattering vector magnitude range $0.025 \ldots 0.18\text{Å}^{-1}$. Neutron wavelength of 10 and 17 Å was used for the range of spin echo times up to 50 ns and 250 ns, respectively. NSE relaxation was measured for both free protein filament and the presynaptic complex with single-stranded phage M13 DNA.

The relaxation curves could be fitted to a single-exponential decay model. The effective diffusion coefficient of the protein self-polymer seemed to increase over 3-fold near the structure factor minimum (0.07Å). The effective diffusion coefficient of the presynaptic complex was lower and did not exhibit the same marked dependency on the scattering angle. Qualitatively similar results (with the quantitative differences consistent with the changes in the solvent viscosity) were obtained in 14 C and 28 C measurements, showing no sign of a transition in the dynamics properties of the protein in this temperature range. The effective diffusion of the filaments observed was in both cases much higher

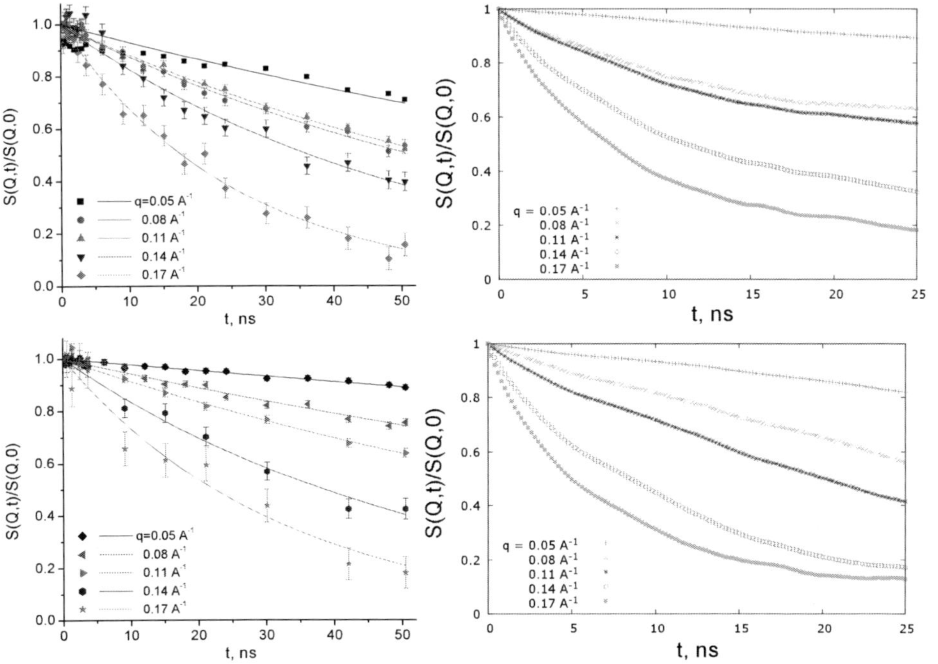

Figure 5. NSE experimental (left) and model (right) relaxation curves for RecA from D.radiodurans (top) and its presinaptic complex (right).

than would be expected for the rigid filament (that were over 100 nm in length, as confirmed by SANS measurements on KWS-2 and KWS-3 spectrometers), and, in the case of the selfpolymer, was similar to the simulated for a translational diffusion of RecA dodecamer (ca. 20 nm in length, figure 5 and 6).

6. Conclusion

Using molecular modeling and its verification via SANS experiments its possible to build realistic all-atom models of quite large macromolecular multicomponent complexes like it was done for RecA::RecX::ssDNA and RecX::ssDNA complexes (this structure was predicted via MM and MD and then verified by SANS).

The molecular dynamics model of RecA filament yields some valuable predictions about the large-scale motion of the filament, particularly regarding the

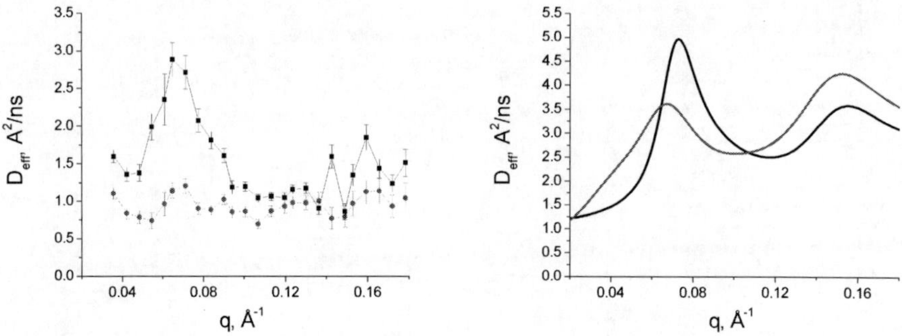

Figure 6. Diffusion for experimental NSE data (left) and model (right) for RecA from *D.radiodurans* and its presinaptic complex.

filament flexibility and its helical pitch (that is in solution decreases from 64 Å (crystallographic data) down to 60 Å in self-polymer and increases to ca.80 Å in the presynaptic complex) and is generally in agreement with the experimental small-angle scattering and NSE data. Spin echo experiment reveals large-scale dynamics in the protein filament, which, in the framework of the MD modeling, should be assigned to large-scale protein bending motion. Formation of the presynaptic complex significantly reduced the filament mobility.

Acknowledgments

This work was founded by RFBR grants 14-04-00817-a and 14-24-01103-ofi-m. The results of the work were obtained using computational resources of MCC NRC ¡ ¡Kurchatov Institute¿ ¿
(http://computing.kiae.ru). We also extend gratitude to Aurel Radulescu (JCNS at FRM2, Munich, Germany), Alexander Kuklin (FLNP JINR, Dubna, Russia) and Peter Falush (ILL, Grenoble, France) for their part in managing neutron scattering experiments at the respective institutions. Special thanks to Dima Baitin (PNPI NRC KI, Gatchina, Russia) and Daria Chervyakova (PNPI NRC KI, Gatchina, Russia) for biological samples preparations.

References

[1] A. I. Roca and M. M. Cox. The RecA protein: structure and function. *Crit Rev Biochem Mol Biol*, 25(6):415–456, 1990.

[2] Randall M. Story, Irene T. Weber, and Thomas A. Steitz. The structure of the e. coli reca protein monomer and polymer. *Nature*, 355(6358):318–325, 1992.

[3] Zhucheng Chen, Haijuan Yang, and Nikola P. Pavletich. Mechanism of homologous recombination from the reca-ssdna/dsdna structures. *Nature*, 453(7194):489–494, 2008.

[4] Margaret S. VanLoock, Xiong Yu, Shixin Yang, Vitold E. Galkin, Hao Huang, Shyamala S. Rajan, Wayne F. Anderson, Elizabeth A. Stohl, H.Steven Seifert, and Edward H. Egelman. Complexes of reca with lexa and recx differentiate between active and inactive reca nucleoprotein filaments. *Journal of Molecular Biology*, 333(2):345 – 354, 2003.

[5] Elisabeth DiCapua, Manfred Schnarr, Rob W.H. Ruigrok, Peter Lindner, and Peter A. Timmins. Complexes of reca protein in solution: A study by small angle neutron scattering. *Journal of Molecular Biology*, 214(2):557 – 570, 1990.

[6] P.A. Timmins, R.W.H. Ruigrok, and E. DiCapua. The solution structure of reca filaments by small angle neutron scattering. *Biochimie*, 73(2..3):227 – 230, 1991.

[7] D.V. Lebedev, D.M. Baitin, A.Kh. Islamov, A.I. Kuklin, V.Kh. Shalguev, V.A. Lanzov, and V.V. Isaev-Ivanov. Analytical model for determination of parameters of helical structures in solution by small angle scattering: comparison of reca structures by {SANS}. {*FEBS*} *Letters*, 537(1-3):182 – 186, 2003.

[8] Michael Petukhov, Dmitry Lebedev, Valery Shalguev, Akhmed Islamov, Aleksandr Kuklin, Vladislav Lanzov, and Vladimir Isaev-Ivanov. Conformational flexibility of reca protein filament: Transitions between compressed and stretched states. *Proteins: Structure, Function, and Bioinformatics*, 65(2):296–304, 2006.

[9] Yu Garmay, A Shvetsov, D Karelov, D Lebedev, A Radulescu, M Petukhov, and V Isaev-Ivanov. Correlated motion of protein subdomains and large-scale conformational flexibility of reca protein filament. *Journal of Physics: Conference Series*, 340(1):012094, 2012.

[10] Rakhi Rajan and Charles E. Bell. Crystal structure of reca from deinococcus radiodurans: Insights into the structural basis of extreme radioresistance. *Journal of Molecular Biology*, 344(4):951–963, 2004.

[11] Ruben Abagyan, Maxim Totrov, and Dmitry Kuznetsov. Icm: A new method for protein modeling and design: Applications to docking and structure prediction from the distorted native conformation. *Journal of Computational Chemistry*, 15(5):488–506, 1994.

[12] Ruben Abagyan and Maxim Totrov. Biased probability monte carlo conformational searches and electrostatic calculations for peptides and proteins. *Journal of Molecular Biology*, 235(3):983 – 1002, 1994.

[13] Stefania Ragone, Joseph D. Maman, Nicholas Furnham, and Luca Pellegrini. Structural basis for inhibition of homologous recombination by the recx protein. *EMBO J*, 27(16):2259–2269, 2008.

[14] Berk Hess, Carsten Kutzner, David van der Spoel, and Erik Lindahl. Gromacs 4: Algorithms for highly efficient, load-balanced, and scalable molecular simulation. *Journal of Chemical Theory and Computation*, 4(3):435–447, 2008.

[15] A.V. Shvetsov, A.E. Shmidt, D.V. Lebedev, and V.V. Isaev-Ivanov. Method for calculating small angle neutron scattering spectra using all atom molecular dynamics trajectories. *Journal of Surface Investigation. X ray, Synchrotron and Neutron Techniques*, 7(6):1124–1127, 2013.

[16] A. Dudkina, A. Schvetsov, I. Bakhlanova, and D. Baitin. Change of filamentation dynamics of reca protein induced by d112r amino acid substitution or atp to datp replacement results in filament resistance to recx protein action. *Molecular Biology*, 45:500–507, 2011.

[17] D. I. Svergun, S. Richard, M. H. J. Koch, Z. Sayers, S. Kuprin, and G. Zaccai. Protein hydration in solution: Experimental observation by x-ray and neutron scattering. *Proceedings of the National Academy of Sciences*, 95(5):2267–2272, 1998.

[18] Alexey V. Shvetsov, Dmitry V. Lebedev, Daria B. Chervyakova, Irina V. Bakhlanova, Igor A. Yung, Aurel Radulescu, Aleksandr I. Kuklin, Dmitry M. Baitin, and Vladimir V. Isaev-Ivanov. Structure of recx protein complex with the presynaptic reca filament: Molecular dynamics simulations and small angle neutron scattering. *FEBS Letters*, 588:948 – 955, Jan-03-2014 2014.

In: Computational Materials ... ISBN: 978-1-63482-541-2
Editor: K. T. Kholmurodov, pp. 25-34

Chapter 3

Oleic Acid in Benzene and Decalin Solutions: Solvation Shell Effect on Small-Angle Neutron Scattering

***Roman A. Eremin*[1,*], *Kholmirzo T. Kholmurodov*[1,2], *Viktor I. Petrenko*[1,3] and *Mikhail V. Avdeev*[1]**
[1]Joint Institute for Nuclear Research,
Dubna, Moscow Region, Russia
[2]International University Dubna,
Dubna, Moscow Region, Russia
[3]Taras Shevchenko National University,
Kyiv, Ukraine

Abstract

The limiting solutions of oleic acid (OA) in deuterated benzene and decalin were previously studied by molecular dynamics (MD) simulations [1]. The spatial neutron scattering length density distributions in the solute-solvent interface area obtained in [1] are refined in the scope of current contribution. In addition, the influence of the observed anisotropic solvation shells on small-angle neutron scattering (SANS) is analyzed. In turn, SANS is used widely for organic solutions internal structure/interaction investigations. Thus, the developed approach of MD

*E-mail address: era@jinr.ru

simulation data use for SANS signal modeling allows to perform the comprehensive scattering data interpretation in case of small solutes as well as strongly anisotropic ones.

Keywords: Solute-Solvent Interaction, Small-Angle Neutron Scattering, Oleic Acid, Deuterated Benzene, Deuterated Decalin

1. Introduction

Small-angle neutron (SANS) and X-ray scattering (SAXS) are powerful techniques in the field of condensed matter structural investigations for systems include inhomogeneities (scatters) with size from 1 to 100 nm (molecular solutions, colloids, porous materials, etc.). Due to the facts that neutrons are scattered by nuclei and different isotopes of one element have different scattering properties (scattering lengths), there is a possibility to study organic solutions by means of SANS using hydrogen/deuterium substitution. For example, in [2] the solutions of mono-carboxylic acids (which are used widely as surfactants within colloidal system stabilization, e.g., ferrofluids) in deuterated (d-) benzene were studied in order to determine the peculiarities of acid-acid interaction. The shape-independent homogeneous Guinier approximation was used during the scattering data analysis in [2]. On the other hand, the specific solvent ordering in the vicinity of solute surface is able to influence SANS signal especially for small solutes (alkyl radical characteristic cross-section diameter is 0.4 nm and lies below the method resolution limit). For the mentioned reason, the question whether the anisotropic solvation shell considerably effects SANS for mono-carboxylic acids solutions in benzene and more complex solvents is open.

Nowadays, molecular dynamics (MD) modeling techniques are applied widely in different fields of condensed matter physics as well as biological research [3]. The computer simulations approaches provide information on the systems consist of several thousand atoms. The most important advantage of MD approach is the detailed dynamical information about each atom (position, velocity, etc.) in the studied system that helps within experimental data interpretation. As an example, previously MD techniques were used for microstructural interpretation of SAXS [4, 5] and SANS [6, 7] experimental results.

Our recent investigations [8, 9] are related to saturated mono-carboxylic (namely, myristic and stearic) acids organic solutions. For them the scattering length density (SLD) spatial distributions obtained in MD simulations are close to cylindrical in assumptions of acid molecules all-*trans* rigid structure [3](Chapter 10). For this reason, a cylindrical scattering particle model could be applied, and significant solvation shell influence on scattering was proven. Recent investigation [1] demonstrated that in case of unsaturated oleic acid (OA) in d-solvent solutions the obtained SLD distributions were non-uniform in the vicinity of acid molecule. The obtained in [1] solvent SLD values in the inner and outer part of the acid molecule bent were found to be considerably different. The aims of the current wort are a) MD-based SANS by OA solutions modelling and b) elucidating of the obtained SLD distributions properties effect on that.

2. SANS Data Modeling by MD Data

2.1. MD Simulation Details

The spatial discretization step was halved in comparison with the previously used one [1]; OA molecule was fixed at the coordinate system origin for simplicity; the SLD calculation/averaging time period was increased up to 4 ns. The simulation MD cell sizes were approximately $6 \times 6 \times 6$ nm^3 both for benzene- and decalin-based solutions. All calculations were performed with DL_POLY 2 general-purpose code [11]. The OA molecule as well as the both studied d-solvent ones were considered to have rigid structure as it was assumed previously for straight saturated myristic and stearic acid [8, 9]. Other MD simulations details were described in the mentioned works [1, 8] comprehensively.
As the result, MD trajectories files (HYSTORY) were used for the spatial SLD distributions calculations using method reported in [3](Chapter 10) with a cubic elementary unit (cube edge $a = 0.05$ nm). The resulting longitudinal and transversal SLD contrast (local SLD value after that of neat solvent, ρs, subtraction) dependencies are shown in Figure 1 (right and left, respectively) for OA in d-benzene (Figure 1, top) and d-decalin (Figure 1, bottom) solutions.

2.2. SANS Data Modeling

As it can be seen from Figure 1, the obtained spatial SLD distributions are not uniform due to the bent in the OA molecule structure (differences of SLD contrast values in the inner bent areas in comparison with the outer ones) both

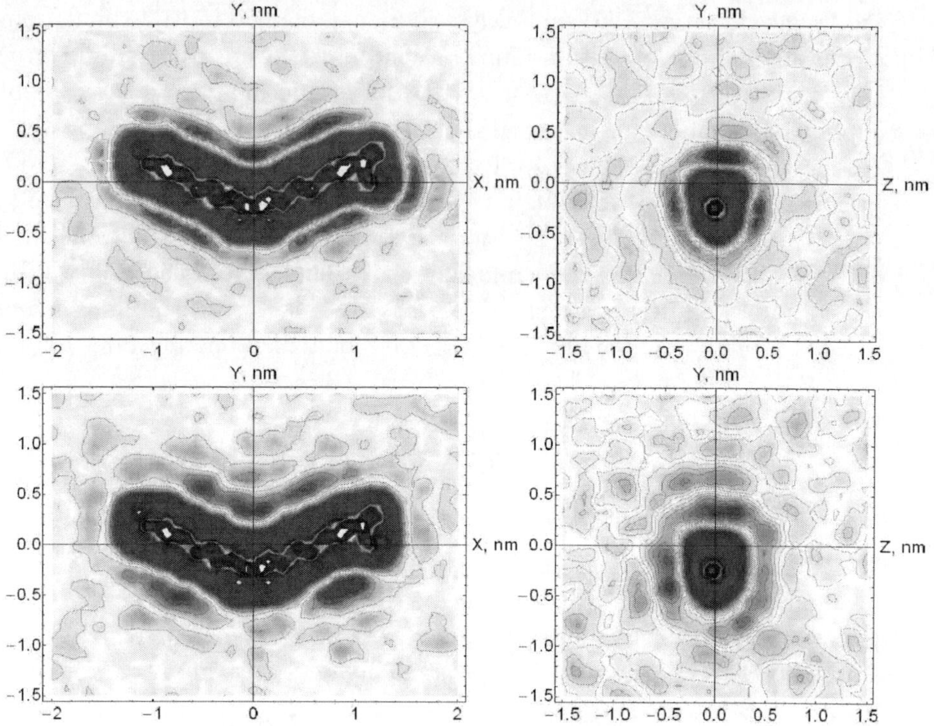

Figure 1. The longitudinal (left) and transversal (right) neutron SLD contrast distributions for OA in deuterated benzene (top) and decalin (bottom) solutions. The following coloring map is used: red and green – negative and positive SLD contrast areas, respectively, white contrast is approximately zero. The 3rd order spline function is used for smoothing.

for d-benzene- and d-decalin-based solutions. The OA molecule has no simple intrinsic symmetry in contrast to the saturated acid molecules. For this reason, the SANS data calculations in the scope of current research were performed basing on the cube method [10]. The obtained SLD spatial distributions can be directly used for SANS modelling according to the following formula for scattering amplitude value [10]:

$$A(\vec{q}) = \sum_i \rho_i \frac{\sin aq_x}{aq_x} \frac{\sin aq_y}{aq_y} \frac{\sin aq_z}{aq_z} \exp i\vec{q}\vec{R}_i \qquad (2.1)$$

where $\vec{q} = \{q_x, q_y, q_z\}$ has a meaning of the transferred wave vector, $\vec{R}_i$ and ρ_i are center position of ith elementary cube and its SLD contrast value known from MD data. After summation over the required number of elementary cubic volumes in the vicinity of OA molecule for a fixed $\vec{q}$ the one can obtain $A(\vec{q})$ value. In order to find $A(q)$ value (that squared determines experimentally observable scattering intensity) a number of $\vec{q}$ spread uniformly over the sphere with radius q should be considered with subsequent averaging over the whole solid angle in the reciprocal space. The described procedure is equivalent to the amplitude average over the all possible orientations of non-spherical scattering particle in solution. As a set of the mentioned points the sixty vertexes coordinates of a truncated icosahedron (vertexes of a soccer ball or carbon atom positions in fullerene C_{60} molecule) were used. The resulting scattering curves (scattering intensities via q dependencies) were evaluated in the following way [10]:

$$I(q) = \mid A(q) \mid^2 \tag{2.2}$$

for eight q values (multiple to the number of processor cores used in parallel evaluations) spread uniformly in the range from 0.1 to 5.0 nm^{-1}. The corresponding to Eqs.1 and 2 *Mathematica* program is presented in Listing 1.

Listing 1

```
S[t_]: = Sin[t]/t; q = Table[q, q, 0.01, 0.51, 1/7 (.51-.01)];
rot = PolyhedronData["TruncatedIcosahedron","VertexCoordinates"];
SLD = Flatten[ParallelTable[{LAY[[i, j, 1]], LAY[[i, j, 2]]−ρs},
{i, Length[LAY]}, {j, Length[LAY[[i]]]}], 1];
AbsoluteTiming[A = ParallelTable[
Abs[Sum[10^10 SLD[[i, 2]] step^3 10^-24 Product[S[(q[[qn]] rot[[rn]])[[j]]], {j, 3}]
Exp[I (q[[qn]] rot[[rn]]).SLD[[i, 1]]],
{i, Length[SLD]}]], {rn, Length[rot]}, {qn, Length[q]}]];
curve1 = Mean[A^2]; curve1=Table[q[[qn]], curve1[[qn]], {qn, Length[q]}];
str = OpenWrite["path1/file1"]; Write[str, curve1]; Close[str]
```

In Listing 1 the auxiliary function S, q values set (in $\AA^{-1}$ for MD data compatibility) and $\vec{q}$ vector orientations (table 'rot', from "rotations") are determined. The 'SLD' table has a meaning of SLD map in the vicinity of acid molecule divided into equidistant layers 'LAY' with respect to the distance to

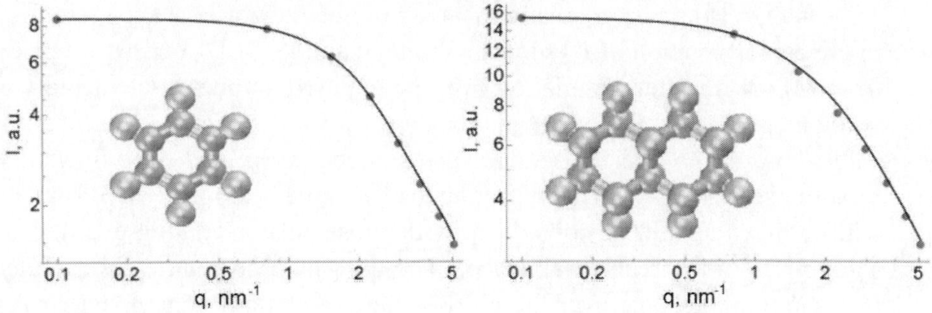

Figure 2. Model SANS curves comparison for OA in deuterated benzene (left) and decalin (right) solutions without (dots) and with (solid curves) equdistant layer averaging (1 a.u. $= 10^{-22}$ cm^{-1})

the acid molecule nearest atom. Such kind division is meaningful for the following calculations and will be described in the next section. The AbsoluteTiming[] command serves for elapsed time determination.
Starting from 'A' value initialization the SANS calculation is performing. The ParallelTable[] function serves to collect $A(q)$ values for all momentum transfer module values in combination with the elements of 'rot' table (totally, $8 \times 60 = 480$). As a result, the SANS curve 'curve1' can be obtained through the averaging over the all $\vec{q}$ orientations by the Mean[] function in accordance with Eq.2 and saved as an untyped file ("path1/file1") for the following use. The obtained SANS curves are shown in Figure 2 (dots) for OA solutions in deuterated benzene and decalin with the unit concentration (1 molecule per 1 cm^3) in arbitrary units (1 a.u. $= 10^{-22}$ cm^{-1}).

3. Quasi-Isotropic Solvation Shells Effect on SANS

In the current section the method of the obtained spatial SLD distributions averaging over equidistant (with respect to OA molecule atoms) layers is proposed. In Listing 2 there are three main parts – the acid molecule atom positions reading, division of the calculated maps into layers and spatial SLD distributions averaging over each layer. Obviously, the subsequent SANS curves calculation can be carried out with according to the Listing 1 program.

Listing 2

```
(*Part 1. Acid molecule atom positions reading*)
ap = .01 Import["acid.pdb", "VertexCoordinates"];
(*Part 2. Division of the calculated maps into layers*)
solvent = "ben"; Edge = 60; step = .5; laystep = .5;
str = OpenRead[solvent<>"-MAP"]; MAP = Read[str]; Close[str];
AbsoluteTiming[
spatial=Flatten[ParallelTable[{{i step - Edge/2, j step - Edge/2, k step - Edge/2},
MAP[[i, j, k]]}, {i, Edge/step}, {j, Edge/step}, {k, Edge/step}], 2];
LAY = ParallelTable[Select[spatial,
dist−step≤Norm[#[[1]]−Nearest[ap, #[[1]], 1][[1]]]≤dist &],
{dist, laystep, 20, laystep}];]
str = OpenWrite["path2/LAY"]; Write[str, LAY]; Close[str];
(*Part 3. Spatial SLD disteributions averaging over each layer*)
PROF = ParallelTable[{l laystep, Mean[Table[LAY[[l, i, 2]]−ρs,
{i, Length[LAY[[l]]]}]]}, {l,Length[LAY]}]
SLD=Flatten[ParallelTable[{LAY[[i, j, 1]], PROF[[i, 2]]},
{i, Length[LAY]}, {j,Length[LAY[[i]]]}], 1];
```

The fixed acid molecule position allows us to get coordinates of its atoms from the corresponding "acid.pdb" file directly. Then, it is necessary to read a SLD map for the chosen solvent type (benzene in Listing 2) from the corresponding file. A special part of code performs division of the map table into subtables in accordance with the distance 'dist' of each map element to the nearest acid molecule atom. It uses the Select[] and Nearest[] commands with step size of 0.05 nm (Å in Listing 2) and saves the obtained distributions as file (as an example, "path2/LAY"). The Listing 2 last part performs averaging of spatial distributions over each layers using auxiliary table 'PROF' (from 'profile'), and flattens the obtained 'SLD'. The resulting spatial distributions are shown for OA in deuterated benzene and decalin solutions in Figure 3. The corresponding SANS curves calculated as it was described above are presented in Figure 2 by solid lines (3rd order interpolation function is used for merging).

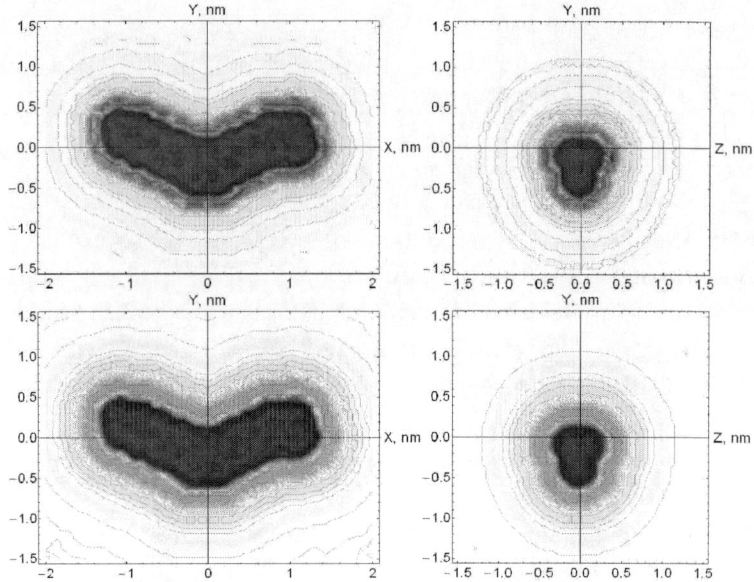

Figure 3. The averaged over equidistant layers longitudinal (left) and transversal (right) neutron SLD contrast distributions for OA in deuterated benzene (top) and decalin (bottom) solutions. Coloring map is remained (see Figure 1 footnote).

4. Conclusion

As it can be seen from Figure 2, the obtained anisotropic behavior of SLD distributions in the vicinity of acid molecules both for d-benzene and d-decalin have no significant effect on SANS. This observation has an important consequence. Due to conformational *trans/gauche* mobility of alkyl residuals in real solutions it is possible that acid molecule has several bents (each appr. 107 deg.). In the scope of current research it is shown that for such conformers the effective scattering signal can be evaluated in approximation of quasi-isotropic solvation shells neglecting the differences of SLD values in the inner and outer regions of molecule bent.

The MD simulations have been performed using the computer software, hardware facilities, and cluster machines at the Central Information and Computer Complex of Laboratory of Informational Technologies at the JINR.

Acknowledgments

The work was partially supported by RFBR (Russian Foundation for Basic Research) project no. 12-02-00649-a, 14-22-01054-ofi_m and joint JSPS (Japan Society for the Promotion of Science) and RFBR project no. 13-04-92100.

References

[1] Eremin R. A., Kholmurodov Kh. T., Petrenko V. I., Avdeev M. V. Solute-solvent interaction in nonpolar solutions of oleic acid as revealed by molecular dynamics simulation, *J. Surf. Investig. X-ray, Synchr. and Neutr. Techn.* 7, 1128-1132 (2013).

[2] Petrenko V. I., Avdeev M. V., Almásy L. *et al.* Interaction of mono-carboxylic acids in benzene studied by small-angle neutron scattering, *Coll. Surf. A* 337, 91-95 (2009).

[3] Kholmurodov Kh. T. (Editor) *Models in bioscience and materials research: molecular dynamics and related techniques*. NY: Nova Sci. Publish. (2013).

[4] Oroguchi T., Ikeguchi M. Effects of ionic strength on SAXS data for proteins revealed by molecular dynamics simulations, *J. Chem. Phys.* 134, 025102 (2011).

[5] Oroguchi T., Ikeguchi M. MD-SAXS method with nonspherical boundaries, *Chem. Phys. Lett.* 541, 117-121 (2012).

[6] Avdeev M. V., Tropin T. V., Bodnarchuk I. A. *et al.* On structural features of fullerene C60 dissolved in carbon disulfide: Complementary study by small-angle neutron scattering and molecular dynamic simulations, *J. Chem. Phys.* 132, 164515 (2010).

[7] Milner E. M., Skipper N. T., Howard C. A. *et al.* Structure and Morphology of Charged Graphene Platelets in Solution by Small-Angle Neutron Scattering, *J. Am. Chem. Soc.* 134, 8302-8305 (2012).

[8] Eremin R. A., Kholmurodov Kh. T., Petrenko V. I., Rosta L., Avdeev M. V. Effect of the solute-solvent interface on small-angle neutron scattering from organic solutions of short alkyl chain molecules as revealed by molecular dynamics simulation, *J. Appl. Cryst.* 46, 372-378 (2013).

[9] Eremin R. A., Kholmurodov Kh. T., Petrenko V. I., Rosta L., Avdeev M. V. Molecular dynamics simulation analysis of small-angle neutron scattering by a solution of stearic acid in benzene, *Phys. Sol. State* 56, 81-85 (2014).

[10] Feigin, L. A., Svergun, D. I. (Editors) *Structure Analysis by Small-Angle X-ray and Neutron Scattering.* Plenum Press, New York (1987).

[11] Smith W., Forester T. R. DL_POLY_2.0: A general-purpose parallel molecular dynamics simulation package, *J. Mol. Graph.* 14, 136-141 (1996).

In: Computational Materials ...
Editor: K. T. Kholmurodov, pp. 35-44
ISBN: 978-1-63482-541-2

Chapter 4

COHERENT CONTROL OF ULTRAFAST REVERSIBLE PHOTOREACTION OF RHODOPSIN

***Olga Smitienko*[1], *Victor Nadtochenko*[1,2,3], *Tatiana Feldman*[1,4,*], *Maria Balatskaya*[5], *Ivan Shelaev*[2], *Fedor Gostev*[2], *Oleg Sarkisov*[2] and *Mikhail Ostrovsky*[1,4]**

[1]Emanuel Institute of Biochemical Physics, Russian Academy of Sciences, Moscow, Russia
[2]Semenov Institute of Chemical Physics, Russian Academy of Sciences, Moscow, Russia
[3]Institute of Problems of Chemical Physics, Russian Academy of Sciences, Chernogolovka, Moscow region, Russia
[4]Biological Faculty, Lomonosov Moscow State University, Moscow, Russia
[5]Faculty of Basic Medicine, Lomonosov Moscow State University, Moscow, Russia

Abstract

Ultrafast reverse photoreaction of visual pigment rhodopsin in the femtosecond time range at room temperature is demonstrated. Femtosecond two pump-probe experiments with time resolution of 25 fs have been

*E-mail address: feldman@mail.ru

performed. The first ump pulse at 500 nm initiated cis-trans photoisomerization of rhodopsin chromophore, 11-*cis* retinal, which resulted in formation of primary ground-state photoproduct within a mere 200 fs. The second pump pulse at 620 nm with varying delay of 200 to 3750 fs relative to the first ump pulse initiated the reverse phototransition of the primary photoproduct to rhodopsin. The results of this photoconversion have been observed on the differential spectra obtained after action of two pump pulses at the time delay of 100 ps. It was found that optical density decreased at 560 nm in the spectral region of bathorhodopsin absorption, and increased at 480 nm where rhodopsin absorbs. Rhodopsin photoswitching efficiency shows oscillations as a function of the time delay between two ump pulses. Molecular mechanism of ultrafast reversible photoreaction of visual pigment rhodopsin may be used as a concept for development of ultrafast optical molecular switch.

Keywords: Rhodopsin; Femtosecond Spectroscopy; Coherent Reaction; Photoreversibility; Molecular Switches

1. Introduction

Visual pigment rhodopsin belongs to a group of photosensitive transmembrane proteins capable of converting energy of light to perform biological functions of the living organisms. It provides light perception as an information carrier. Rhodopsin possesses unique photochemical properties. In these natural pigment-protein complex, chromophore photoisometization rate shows a femtosecond time range [1–6]. Therefore, ever-increasing interest to detailed study of photochemical transformation mechanisms of rhodopsin is observed due to the purpose of using functioning principles of this natural photosensor in creation of fast response devices [7–12].

Bovine rhodopsin is one of the well-studied representatives of retinal-binding proteins. It is typical member of the G-protein-coupled receptor family. Rhodopsin molecule consists of opsin protein and a chromophore, 11-*cis* retinal. Light quantum absorption leads to 11-*cis* retinal isomerization to all-*trans* form. Chromophore photoisomerization provides conformational changes in the protein part of the molecule resulting in formation of intermediates with various lifetimes and spectral properties. Finally, rhodopsin phototransformation (photolysis) results in hydrolysis of Schiff base linkage and all-trans retinal release from the rhodopsin molecule.

Elementary act of chromophore photoisomerization in rhodopsin proceeds during time of about 80-100 fs [2,6,13] with quantum yield of 0.65 [14]. Meanwhile, almost all energy of the light quantum absorbed (¿ 60%) is converted to chemical energy conformational protein rearrangements [15]. Retinal obtains unique photochemical properties of this kind due to its protein environment. It is worth noting that in absorption kinetic curves of primary rhodopsin photoproduct oscillations provided by formation of coherent oscillation wave packets are observed [2–4,13,16].

All rhodopsin photolysis intermediates possess photochromic properties. Absorption of the second light quantum by photolysis intermediate may cause reverse *trans-cis* isomerization of retinal and formation of rhodopsin with 11-*cis* retinal (photoregeneration process) with an insignificant admixture of isorhodopsin with 9-*cis* retinal [17].

Photochromism of rhodopsin, as well as all other retinal-binding proteins, i.e. the ability to exist in two or more stable (quasi-stable) forms, between which reversible phototransitions are possible, allows consideration of these pigment-protein complexes to be prototypes of optical molecular switches [7–9,11,18,19]. For that matter, early stages of retinal-binding proteins phototransformations are the most attractive. Photochromism in bacteriorhodopsin has found use in creation of photosensitive media in optical devices for information writing, transmission and storage with typical switching time of about milliseconds or nanoseconds [20–22]. Yan et al. [23] has described primary experiments on ultrafast reversible photoswitching of bovine rhodopsin affected by femtosecond (300 fs) laser pulses.

The aim of this work was to study photochromic transition of primary intermediates of bovine rhodopsin photolysis containing chromophore in trans-configuration to the ground state of rhodopsin with 11-cis retinal in femto- and picosecond time range. In the work, efficiency dependence for such photochromic transition on time delay between the first and the second pump-pulses was demonstrated.

2. Results and Discussion

The photoexcitation of rhodopsin (Rh_{498}) by 25 fs pump pulse at 500 nm results in 11-cis retinal chromophore photoisomerization to the *trans*-form and formation of primary ground-state rhodopsin photoproduct ($Photo_{570}$) by 200th fs that, following picosecond vibrational relaxation, leads to production of the

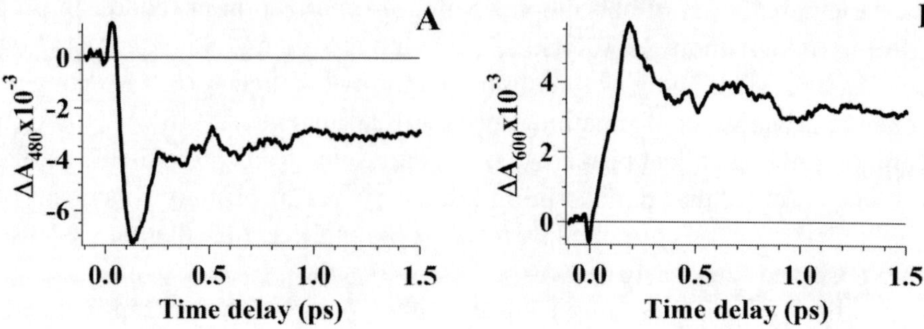

Figure 1. Transient photo-induced absorption kinetic curves of rhodopsin. The kinetic curves were documented by one pump-probe experiments with the pumping by 25 fs pulse at 500 nm and probing at 480 nm (A) and 600 nm (B).

metastable intermediate bathorhodopsin ($Batho_{535}$). Figure 1 shows kinetic curves of photo-induced Rh_{498} absorption at 480 nm (A) and at 600 nm (B).

At the probe wavelength of 480 nm (Figure 1) at the earliest times (up to 50 fs), a low positive absorption associated with excited state absorption of rhodopsin molecule ($S_1 \rightarrow S_n$) is observed [2–4]. At times up to 130 fs, negative absorption, i.e. bleaching of Rh_{498}, is observed. Thereafter, during several picoseconds partial absorption recovery due to transition of a portion of excited molecules to the initial state of Rh_{498} and $Photo_{570}$ formation and its further relaxation to $Batho_{535}$, which also absorb in this spectral range, are observed. At the probe wavelength of 600 nm (Figure 1), $Photo_{570}$ formation may be observed that completes by 200th fs after photoexcitation.

Irrespective of the probe wavelength, for all kinetic curves oscillation processes were observed. Our earlier works [4,7,24] show results of Fourier-analysis that prove the presence of oscillations in dynamics of photo-induced absorption by reaction products. At times of 0.9 1 ps frequencies of vibrational modes comprised by wave packets of $Photo_{570}$, $Batho_{535}$ and Rh_{498} were determined. In the entire $Photo_{570}$ and $Batho_{535}$ absorption band, the same set of frequencies was obtained: 59, 117, 156, 234, 254, and 351 cm^{-1} with notably predominance of 59 cm^{-1} frequency, whereas in Rh_{498} absorption band similar set of frequencies with slightly lower values was obtained: 39, 59, 136, 195, 234, and 371 cm^{-1} with predominance of 136cm^{-1} frequency. Detected oscillation frequencies correspond to vibrational modes which are active within the elementary act of 11-*cis* retinal isomerization and during recovery of Rh_{498} [4,24]. It is worth noting that by 3 ps approximately, the oscillation relaxation process already completed and no oscillations were observed.

For reverse photoreaction of $Photo_{570}$ or $Batho_{535}$ transformation to Rh_{498}, a femtosecond two pump-probe pulse setup with two pump pulses (hereinafter referred to as *pump1* and *pump2*) and supercontinuum probing pulse was used [7,25]. 25 fs *pump1* at 500 nm initiates the reaction: $11-cis-retinal+h\nu_1 \rightarrow trans-retinal$, and by 200th fs $Photo_{570}$ is formed which relaxes to $Batho_{535}$ with characteristic time of 1-3 ps. Differential spectra showing dynamics of $Batho_{535}$ formation and Rh_{498} disappearing in 100 ps after *pump1* action in the spectral range of 410 700 nm were registered (Figure 2, black curve). 30 fs *pump2* at 620 nm was supplied after 25 fs pump1 at 500 nm with time delay from hundreds of femtoseconds to several picoseconds. The probe pulse time delay relative to pump1 was 100 ps. Figure 2 (grey curve) shows results of two pump pulses affecting the sample and following one another with 200 fs time delay. I.e. in this experiment *pump2* affected $Photo_{570}$. By 100th ps after consecutive action of *pump1* and *pump2* a decrease of $Batho_{535}$ dark formation and simultaneous increase of initial Rh_{498} absorption was observed. This result means that *pump2* induces a photoreaction, in which $Photo_{570}$ transforms back to the Rh_{498}. To put it differently, the reverse *trans-cis* isomerization reaction of retinal residue occurs: $trans - retinal + h\nu_2 \rightarrow 11 - cis - retinal$. It is worth noting that after the effect of *pump2* the form of differential spectra did not changed. This may suggest that reverse phototransition $Photo_{570} \rightarrow Rh_{498}$ gives no side products.

Oscillations observed in kinetic curves of photo-induced chromophore isomerization in Rh_{498} reflect dynamics of $Photo_{570}$ coherent wave packet. To put it differently, these oscillations reflect occurrence of quantum interference during the reaction which results in time dependence of different stationary vibrational states. It is expected that the reverse phototransition $Photo_{570} \rightarrow Rh_{498}$ efficiency may significantly depend upon whether in the maximum or in the minimum of oscillations *pump2* is absorbed. Experiments in which time delay between *pump1* and *pump2* varied so that time of *pump2* arrival was synchronized with the oscillation phase present in absorption signals of $Photo_{570}$. The maximum of differential absorption spectrum that reflects formation of $Batho_{535}$ registered 100 ps after pump1 was assumed as a measure of reverse photoreaction efficiency (Figure 2, black curve).

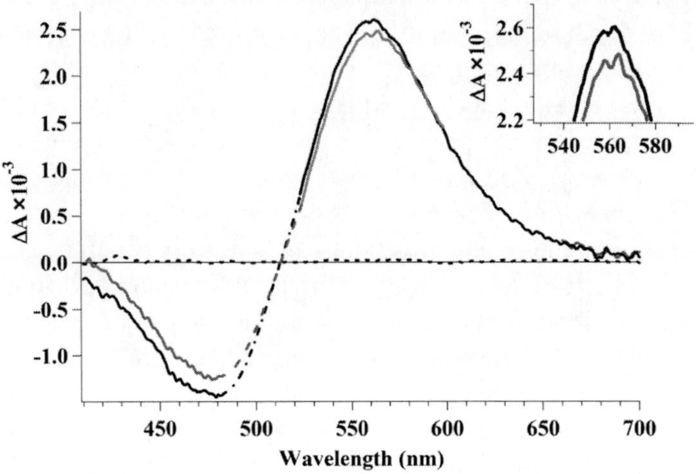

Figure 2. Differential spectra of photo-induced rhodopsin absorption obtained under the effect of one pump pulse of 500 nm (*pump1*) (black curve) and two pump pulses of 500 and 620 nm (*pump1* and *pump2*) (grey curve) that follow with 200 fs time delay and with the probe pulse time delay of 100 ps. For the control, similar spectrum of completely bleaching rhodopsin affected by a single pump pulse of 500 nm (dashed curve) was registered. In spectral regions of pump pulses experimental curves were completed. Insert: expanded part of the $Batho_{535}$ maximum absorption area.

Figure 3 (at the bottom) shows kinetic curve of photo-induced Rh_{498} absorption obtained in the experiment with 25 fs *pump1* at 500 nm, when probed at 620 nm. Figure 3 (at the top) shows values of optical density reflecting variations in absorption of $Batho_{535}$ at 560 nm (circles) and Rh_{498} at 450 nm (triangles) with respect to *pump2* time delay. Optical densities at 450 and 560 nm in the differential spectrum observed 100 ps after exclusive action of *pump1* were taken as a control (Figure 2, black curve). These control values are shown in Figure 3 (at the top) with 0 fs time delay. The results obtained show that the number of $Photo_{570}$ molecules participated in the reverse photoreaction depends upon synchronization of *pump2* time delay with phase characteristics of the wave packet in $Photo_{570}$. In case when *pump2* arrives at the moment of maximum $Photo_{570}$ absorption (200 fs time delay), the reverse photoreaction obtains maximum efficiency. Hence, the differential spectrum recorded 100 ps after excitation (Figure 2, grey curve) shows maximum signal decrease at 560 nm wavelength and maximum signal increase at 450 nm. Photochromic switching is lower, when the control pulse (*pump2*) arrives at the antiphase of coherent vibrational wave packet motion and is absent in the oscillation minimum at 475 fs time delay (Figure 3, at the top). Similar results were obtained in a broad probing spectral range (420-620 nm). Photochromic switching was observed up to 4 ps time delay of *pump2*.

Thus, this work shows, firstly, phototransitions $Rh_{498} \rightarrow Photo_{570} \rightarrow Rh_{498}$ occurring in a femtosecond scale at room temperature. Secondly, it is shown that photochromic switching efficiency depends on the timing of oscillation maximum for the dynamics of the coherent vibrational wave packet of Photo_{570} and *pump2*. Such dependence is the obvious example of coherent control over the reverse photoreaction of Rh_{498} chromophore group photoisomerization.

The results obtained allow the consideration of the visual pigment rhodopsin as a model for creating an ultrafast molecular photoswitch. The mechanism of rhodopsin photoswitching described in this work may underlay the functionality of an optical logical element probably designed for use in optical information transfer and processing devices with a running speed exceeding that of the existing analogues [11]. As photochromic materials, various retinal-binding proteins can be used: visual pigments rhodopsins from invertebrates and vertebrate animal eyes, sensor rhodopsins and photodependant ionic pumps of microorganisms. The use of various mutant and modified protein forms is also possible, modifications being present in both the peptide part of the molecule and chromophore. Therefore, searching for artificial matrices for a retinal chromophore capable of simulating the natural protein matrix is possible. There are some examples for retinal and its analogues [10,2628]. The principle of the coherent control of rhodopsin photochromic reactions induced by femtosecond laser pulses at room temperature proposed in this work and some our other studies [4,7,11] may also be found useful for the development of quantum computers.

3. Experimental Section

Rod outer segments (ROS) from bovine retinas and rhodopsin extracts were prepared by the method of Okada et al. [29]. Transient absorption spectra were measured by the femtosecond pump-supercontinuum probe setup [25]. For investigation of reverse photoreaction and coherent mechanism of direct

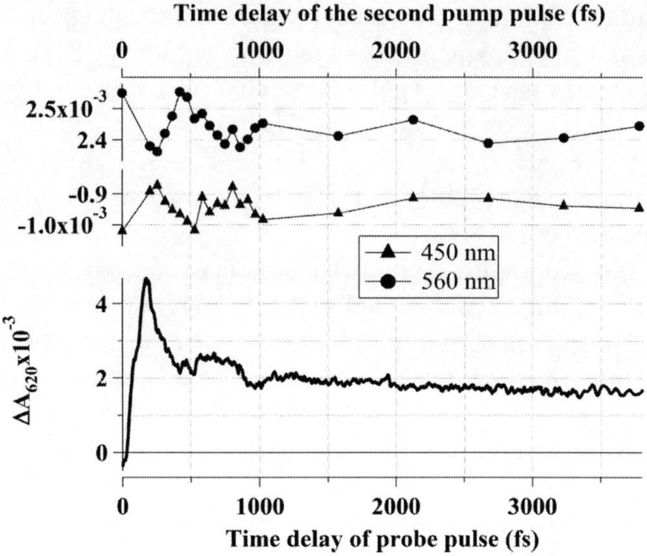

Figure 3. Coherent control of primary ground-state rhodopsin photoproduct phototransition to rhodopsin. (At the bottom) Kinetic curve of photo-induced rhodopsin absorption under the effect of 25 fs *pump1* at 500 nm and probing at 620 nm. (At the top) Variation of photo-induced rhodopsin absorption recorded at probe wavelengths of 450 nm (triangles) and 560 nm (circles) 100 ps after *pump1* and *pump2* excitation at different *pump2* time delays (upper abscissa axis). Optical density at 0 fs time delay was obtained by exclusive rhodopsin excitation by *pump1* and is assumed as a control.

rhodopsin photoreaction two pump pulses and supercontinuum probe pulse were used [7]. All manipulations with samples containing rhodopsin were carried out under dim red illumination.

Acknowledgments

This work is supported by the Russian Academy of Sciences Presidium Program of Fundamental Studies (Programme No. 24) and Russian Foundation for Basic Research No. 12-04-00844-a. In part of the study related to the femtosecond laser spectroscopy the work was supported by The Ministry of education and science of Russia, grant No. 14.604.21.0058 (unique identifier RFMEFI60414X0058).

References

[1] Dobler, J.; Zinth, W.; Kaiser, W.; Oesterhelt, D. *Chem. Phys. Lett.* 1988, 144, 215-220.

[2] Kandori, H.; Shichida, Y.; Yoshizawa, T. *Biophys. J.* 1989, 56, 453-457.

[3] Polli, D.; Alto, P.; Weingart, O.; et al. *Nature* 2010, 467, 440-443.

[4] Schoenlein, R.W.; Peteanu, L.A.; Mathies, R.A.; Shank, C.V. *Science* 1991, 254, 412-415.

[5] Smitienko, O.A.; Mozgovaya, M.N.; Shelaev, I.V.; et al. *Biochemistry* (Moscow) 2010, 75, 25-35.

[6] Taiji, M.; Bryl, K.; Nakagawa, M.; et al. *Photochem. Photobiol.* 1992, 56, 1003-1011.

[7] Yabushita, A.; Kobayashi, T.; Tsuda, M. *J. Phys. Chem. B.* 2012, 116, 19201926.

[8] Mozgovaya, M.N.; Smitienko, O.A.; Shelaev, I.V.; et al. *Dokl. Biochem. Biophys.* 2010, 435, 302-306.

[9] Ostrovsky, M.A.; Weetall, H.H. *Biosensor & Bioelectronics* 1998, 13, 61-65.

[10] Paternolli, C.;Neebe, M.;Stura, E.;et al. *Biomed. Mater. Res. A.* 2009, 88, 947-951.

[11] Tegeder, P. *J. Phys. Condens. Matter.* 2012, 24, 394001.

[12] RU Patent 2,420,773, 10 June 2011.

[13] Weinelt, M.; von Oppen, F. J. *Phys. Condens. Matter.* 2012, 24, 390201.

[14] Nadtochenko, V.A.; Smitienko, O.A.; Feldman, T.B.; et al. *Dokl. Biochem. Biophys.* 2012, 446, 242-246.

[15] Kim, J.E.; Tauber, M.J.; Mathies, R.A. *Biochemistry* 2001, 40, 1377413778.

[16] Cooper, A. *Nature.* 1979, 282, 531-533.

[17] Wang, Q.; Schoenlein, R.W.; Peteanu, L.A.; et al. *Science*. New Ser. 1994, 266, 422-424.

[18] Yoshizawa, T.; Wald, G. *Nature*. 1963, 197, 1279-1286.

[19] Berthoumieu, O.; Patil, A.V.; Xi, W.; et al. *Nano Lett.* 2012, 12, 899903.

[20] Patil, A.V.; Premaraban, T.; Berthoumieu, O.; et al. *Chemistry* 2012, 18, 5632-5636.

[21] US Patent 6,219,160, 17 April 2001.

[22] Birge, R.R. Patent US 5559732, 1996.

[23] US Patent 10,337,348, 18 October, 2005.

[24] Yan, M.; Rothberg, L.; Callender, R. *J. Phys. Chem.* 2001, 105, 856859.

[25] Smitienko, O.A.; Shelaev, I.V.; Gostev, F.E.; et al. *Dokl. Biochem. Biophys.* 2008, 421, 277-281.

[26] Shelaev, I.; Gostev, F.; Mamedov, M.; et al. *BBA Bioenergetics* 2010, 1797, 1410-1420.

[27] Briand, J.; Brm, O.; Rehault, J.; et al. *Phys. Chem. Chem. Phys.* 2010, 12, 3178-3187.

[28] Kandori, H.; Ichioka, T.; Sasaki, M. *Chem. Phys. Lett.* 2002, 354, 251-255.

[29] Sinicropi, A.; Martin, E.; Ryazantsev, M.; et al. *Proc. Natl. Acad. Sci. U. S. A.* 2008, 105, 1764217647.

[30] Okada, T.; Takeda, K.; Kouyama, T. *Photochem. Photobiol.* 1998, 67, 495-499.

In: Computational Materials ...
Editor: K. T. Kholmurodov, pp. 45-51
ISBN: 978-1-63482-541-2

Chapter 5

MECHANISM OF INTERACTION OF MACROMOLECULES WITH SURFACTANT MICELLES IN AQUEOUS SOLUTION: A MOLECULAR DYNAMICS SIMULATION

Aram A. Shahinyan*, Levon H. Arsenyan and Armen H. Poghosyan
International Scientific-Educational Center,
National Academy of Sciences,
Yerevan, Republic of Armenia

Abstract

The mechanism of formation and structural changes of polyvinylalcohol (PVA) and sodium pentadecyl sulfonate ($SPDS$) containing mixed aggregate in aqueous solution were investigated using molecular dynamics (MD) simulation. It was determined, that mixed aggregate was formed due to $SPDS$ separate molecules hydrophobic attraction towards the PVA macromolecule in molecular solution, while in the micellar solution interactions are between the micellar fragments and the PVA macromolecule. These features influenced upon the process of mixed aggregates formation at the initial stage.

Keywords: Micelle, Polymer Coil, Gyration Radius, Mixed Aggregate, Molecular Dynamics Simulation

*E-mail address: artsha@sci.am

1. Introduction

In this paper, the mechanism of mixed polymer-amphiphilic aggregates formation in water was examinned using MD study. Interaction between the macromolecule and micelle of amphiphilic compounds has practical importance, especially from chemical and oil industrial perspectives, as well as in medicine science due to its selectivity in drug delivery [1,2]. In papers [3,4] selective interaction, between anionic amphiphilc compound micelles and molecules of PVA or PVP (polyvinylpyrrolidone) in aqueous solution was shown. For better interpreting the amphiphilic compounds and polymers mixed aggregates formation process, it is important to explore the interaction mechanism between polymer and amphiphilic molecules in solution. For example, a series of studies [3-8] show that there is a selective interaction between micelles of amphiphilic compaunds (for instance $SPDS$ or sodium dodecylsulfate SDS) and polymer molecules (PVA or PVP) in an aqueous solution, and nonlinearity of certain properties of the system depends on the ratio of amphiphilic compounds and polymers concentrations. In this paper, using the MD method, we have investigated the mechanism of formation and structure of PVA-$SPDS$ mixed aggregate.

2. Model Calculation. Results and Discussion

Since PVA is not a polyelectrolyte one can assume that the interaction between PVA molecules and $SPDS$ in water should consist of: a) weak hydrophobic interactions between the polymer (PVA) molecules backbone and the hydrocarbon chains of $SPDS$ molecules; b) dipole-dipole and dipole-ion interactions between hydroxyl groups of the PVA macromolecule and dissociated or non-dissociated polar groups of $SPDS$ molecules; c) hydrogen interactions between hydroxyl groups of the PVA molecule and water molecules. The MD simulation of PVA/$SPDS$/water system was performed using GROMACS software package for the following two cases: the model of previously simulated PVA macromolecule coil was inserted into the $SPDS$ aqueous solution model system, i.e. in this case formation of the micelle occurs at the presence of interaction of $SPDS$ molecules with the polymer molecule; in second case, the model of PVA coil was embedded into the model of $SPDS$ micellar solution, i.e. in this case, the interaction of polymer with a formed micelle was studied. Fig.1 shows the starting configuration of micelle containing 128 $SPDS$ molecules

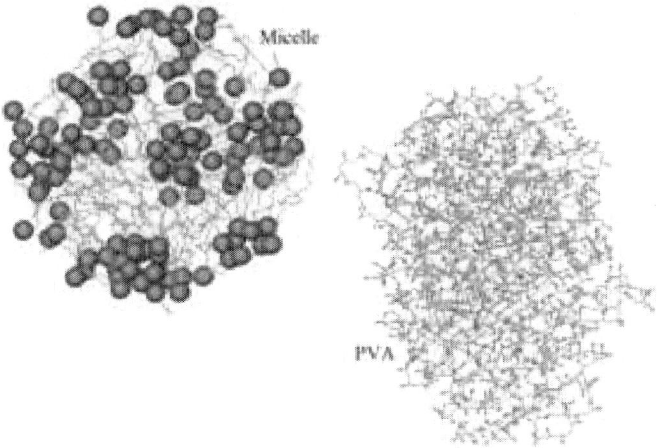

Figure 1. Snapshots of thermodynamically equilibrium model of $SPDS$ micelle and the coil of PVA before the simulation.

and polymer coil in the equilibrium state, before the simulation. The VdW spheres indicate the polar groups of $SPDS$ molecules at the surface of micelle, the gray sticks- the hydrocarbon chains of $SPDS$ molecules. Hydroxyl groups of PVA molecule are colored red. Water molecules are not shown to reduce the volume of the figure.

Fig. 2 shows the polymer coil and the micelle, that have been fitted together after 2ns.

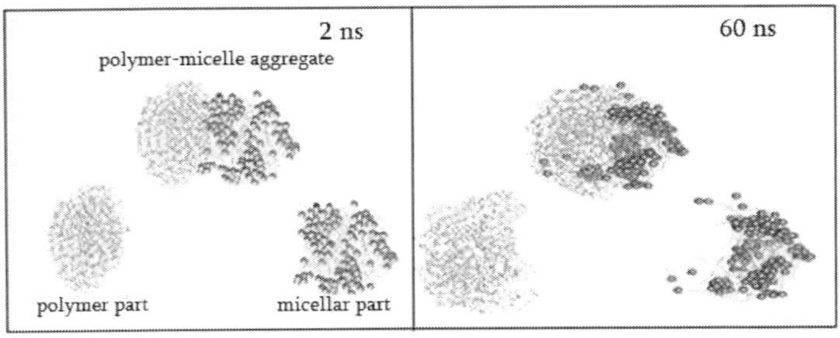

Figure 2. Interaction of $SPDS$ micelle and the coil of PVA after 2 and 60 ns of simulation run. Separately are presented polymeric and micellar parts of aggregate.

The polymeric and micellar parts of the aggregate, as well as the mixed aggregates are displayed separately. As one can see on Fig. 2 and 3, after the the interaction, the micelle and the polymer coil starts to distort the shape, and the polymer coil is deformed and sticking to micelle formation.

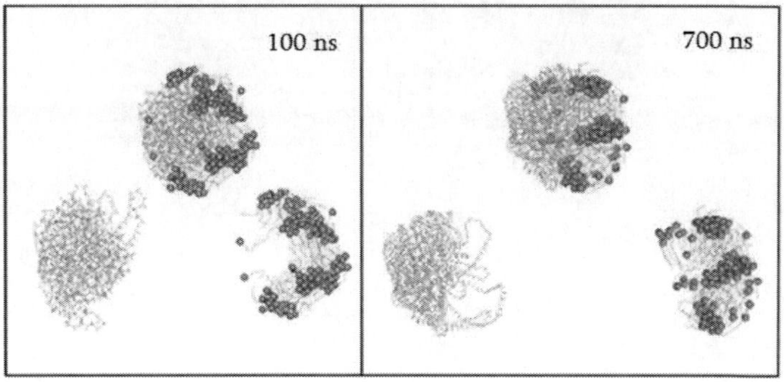

Figure 3. Interaction of $SPDS$ micelle and the coil of PVA after 100 and 700 ns of simulation. Separately are presented polymeric and micellar parts of the aggregate.

It is obvious from obtained data that during interaction, the PVA molecule penetrates into the hydrophobic core of the micelle, therefore it may be assumed that interaction between the PVA molecule and $SPDS$ micelle is mainly of hydrophobic nature and after the mixed aggregate formation, and further increase of simulation time leads to amphiphilic molecules and micelle fragments gradual re-distribution over the surface and volume of the mixed aggregate. Now let us examine formation of mixed aggregate after mixing PVA coil with the molecular solution of $SPDS$ in water (starting from random distribution), before micelles formation. In this case, formation of mixed aggregate has already occurred due to the interaction of polymer with separate amphiphilic molecules. Fig. 4, 5 illustrate the process of mixed aggregate formation in molecular solution of $SPDS$ in presence of PVA coil, before the simulation start and after 20 ns (Fig.4), after 300 ns and 1250 ns (Fig.5) simulation.

It is seen from Figure 4, in parallel with mixed aggregate formation, process of $SPDS$ molecules premicellar associates formation occurs, which is completely adsorbed on the coil of polymer later (Figure 5). Comparing the mixed aggregates obtained from the interaction of PVA with micelles and individual molecules in the molecular solution of $PDSN$ (Figure 2, 3, 4 and 5), we see

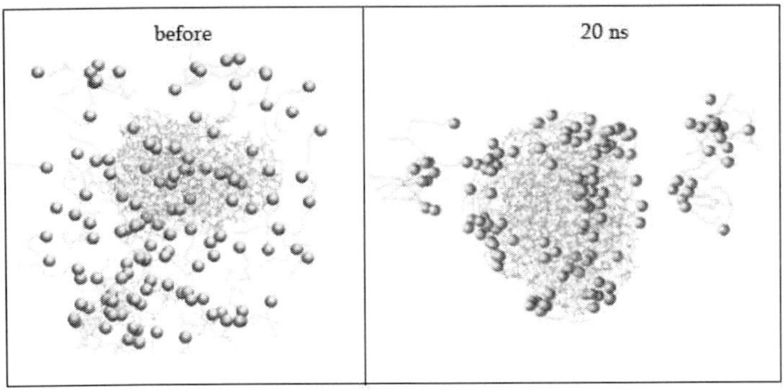

Figure 4. Coil of PVA in the molecular solution of $SPDS$ in water: before the start and after 20 ns of simulation.

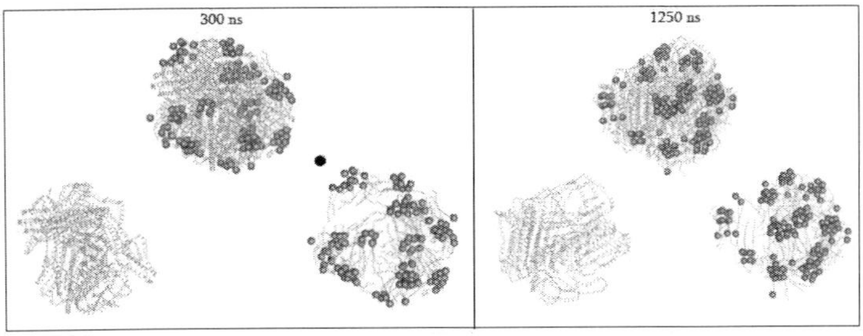

Figure 5. Formation of $PVA/SPDS$ mixed aggregate at the molecular solution of $SPDS$ in the presence of PVA coil after 300 and 1250 ns of simulation.

that in second case, molecules of $PDSN$ are distributed on the surface of mixed aggregate more homogeneous than they are in the case of mixed aggregate formation at micellar solution. Thereby, the formation of mixed aggregate PVA / $SPDS$ occurs faster and deeper in molecular solution than in micellar solution. Probably, in the latter case it is due to the splitting of micelle into separate fragments, i.e. to overcome the forces of the hydrophobic attraction of amphiphilic molecules in a micelle that requires a certain amount of energy. Some important details of polymer / amphiphile mixed aggregates formation can be figured out by investigating the dynamical changes of main parameters describing the mixed aggregate during its formation. Change of the PVA coil average radius of gyration in mixed aggregate depending on the simulation time is shown in Fig. 6.

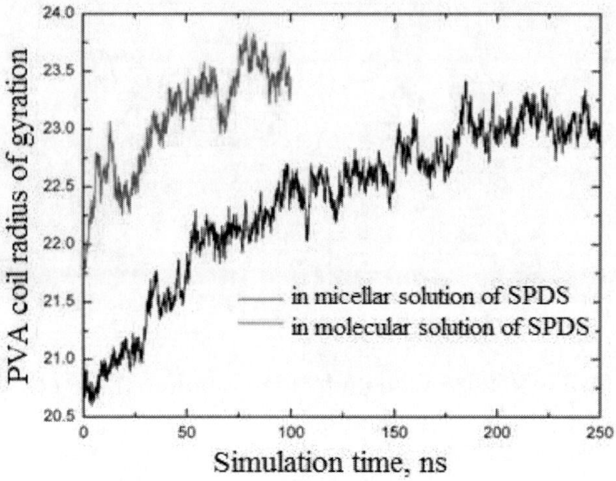

Figure 6. Change of the coil of the PVA radius of gyration during the formation of mixed aggregate in micellar and molecular solutions of $SPDS$.

Fig. 6 shows that the mixed $PVA/SPDS$ aggregate formed in the molecular solution of $SPDS$ has a looser structure than it has in micellar solution. These results are fully consistent with the data presented in the Figures 2-5. It is obvious from Fig.6 that mixed $PVA/SPDS$ aggregate formed in the molecular solution of $SPDS$ has softer structure than it has in micellar solution. Therefore the mixed aggregate formation process initial stage in the molecular solution is more intensive than in the micelar solution due to the dynamical changes average radius of gyration of the polymer coil in the mixed aggregates correlate with the results on the structure of mixed aggregates.

References

[1] Kwak J.C.T., Editor, ”Polymer–Surfactant Systems”, Marcel Dekker (1998) New York. Shahinyan A.A., Khanamiryan L.A., Aivazyan O.M. ”On the influence of conformation of macromolecules and type of amphiphilic compounds on the structure of mixed agregates, simulating lipid-protein hydrophobic complexes”. *Reports of Academy of Sciences of the Armenian SSR*, 83, N2 (1976) 113-116 (in Russian)

[2] Khanamiryan L.A., Shahinyan A.A. "Influence of the electrolyte on the structure of mixed aggregates consisting of the anionic surfactant and water soluble polymer". *Arm. Chem. J.*, 31, N9 (1978) 705-707 (in Russian)

[3] Shahinyan A.A., Minasyants M.Kh., Khanamiryan L.A., Sarkisyan A.G. "X-ray and rheological study of mixed aggregates composed of surfactant and water-soluble polymers". *Proceeding of the Academy of Sci. of the Arm. SSR*, 15, Issue 1 (1980) 55-58 (in Russian).

[4] Minasyants M.Kh., Sarkisyan A.G., Shahinyan A.A., Khanamiryan L.A., Chistiakov I.G. "X-ray diffraction study of the influence of NaCl concentration on the structure of mixed lyotropic liquid crystals consisting of amphiphilic molecules and water-soluble polymer". *Cristallography*, 24, N4 (1979) 864-865 (in Russian)

[5] Nagarajan R. "Thermodynamics of Surfactant-Polymer Interactions in Dilute Aqueous Solutions". *Chemical Physics Letters*, 76 (1980) 282-286.

[6] Shahinyan A.A. "The role of structural organization of ionic micelles at the mechanism of forming macromolecules in emulsions". *Published by Academy of Sciences of Armenian SSR: Yerevan, Armenian SSR*, (1985) (in Russian).

[7] Cabane B. "Condensation of Surfactants Micelles on a polymer Molecule in Aqueus Solution". *Solut. Bevav. Surfactants: Theor. And Applied Aspects Proc. Int. Symp.*, Podstam, 1980, 1 New York, London (1982) 661-664.

[8] Edson Minatti, David P. Norwood, and Wayne F. Reed. "Surfactant/Polymer Assemblies. 2. Polyelectrolyte Properties". *Macromolecules* (1998) 31, 2966-2971.

In: Computational Materials ... ISBN: 978-1-63482-541-2
Editor: K. T. Kholmurodov, pp. 53-62

Chapter 6

QUADRUPOLE EXCITATION IN SPIN TUNNELING IN MAGNETIC NANO-PARTICLES

Khikmat Kh. Muminov* ***and Yousef Yousefi***[†]
Physical-Technical Institute named after S.U. Umarov
Academy of Sciences of Republic of Tajikistan
Dushanbe, Tajikistan

Abstract

Spin tunneling effect in nano-particles Fe_8 and Mn_{12} is studied by instanton calculation technique using the SU(3) generalized spin coherent states as a trial functions. Energy level splitting is calculated by use of the spin coherent states path integrals and its dependence on the excitation of quadrupole dynamics is shown. For both nano-particles, tunnel splitting arises due to presence of the generalized Berry phase in action, which causes interference between tunneling trajectories (instantons). It is established that taking into account of quadrupole excitation changes not only the location of the quenching points, but also the number of these points in the energy level splitting. Consideration of both dipole and quadrupole excitations in the classical energy reduces the number of quenching points to the experimentally observed number of steps in hysteresis loops of these nano-particles.

*E-mail address: muminov@phti.tj, yousof54@yahoo.com
†E-mail address: yousof54@yahoo.com

Keywords: Spin Tunneling, Nano-Particles, Fe_8, Mn_{12}, Quadrupole Excitations

1. Introduction

Today more than 5000 different magnetic molecular nano-clusters are known, and about a 100 of them exhibit some behaviour that is of interest from point of view of future applications [1]. These molecules are intermediate between simple paramagnetic salts such as $CuSO_4.K_2SO_4.6H_2O$ or $Ce_2Mg_3(NO_3)_{12}.24H_2O$, in which the magnetic entities are single transition metal or rare earth ions, and super-paramagnetic particles of sub-micrometer size. These nano-clusters demonstrate a phenomena, which are known as hysteresis at molecular level. There is an opportunity for understanding the energy relaxation process in great detail since the magnetic entities are molecules and thus well defined with identical size, shape, orientation, and also the interactions of the magnetic and nonmagnetic degrees of freedom are relatively well understood. Some of these issues are similar to those that arise in the study of magnetization reversal of small magnetic particles, and it is hoped that the molecular systems will offer insights into the latter, which has applications in magnetic recording and storage technologies.

A single-molecule magnet (SMM) is a molecule that behaves as an individual nanomagnet. Because of their small size and precise characterizability, molecular nanomagnets exhibit many fascinating quantum phenomena, such as macroscopic quantum tunneling of magnetization and Berry-phase interference [2,3]. They straddle the quantum mechanical and classical worlds, residing in a middle ground that is of abiding interest to physicists.

The magnetic core of Mn_{12}-ac, Fig.1, has four Mn^{4+} $(S = 3/2)$ ions in a central tetrahedron surrounded by eight Mn^{3+} $(S = 2)$ ions. The ions are coupled by superexchange interaction through oxygen bridges with the net result that the four inner and eight outer ions point to opposite directions, yielding a total spin $S = 10$ [4].

The magnetic cluster Fe_8, Fig.2, like Mn_{12}-ac, has a spin ground state $S = 10$, which arises from competing antiferromagnetic interactions among the eight $S = 5/2$ Fe spins and the spin dynamics are quite similar. Whereas Mn_{12}-ac has an easy axis and an essentially isotropic hard plane, Fe_8 has three non-equivalent axes [5,6].

A variety of experimental techniques (electron spin resonance, ac susceptibility, magnetic relaxation, Mossbauer spectroscopy, neutron scattering) indicates [6-9] that experimental data are in agreement with the following model Hamiltonian

$$\hat{H}_{Fe} = -k_1 {J_z}^2 + (k_1 - k_2) {J_x}^2 - k_4({J_+}^4 + {J_-}^4) - g\mu_B J.h \qquad (1.1)$$

The parameters g, k_1 and k_2 are obtained through a variety of experimental techniques [10]. with J = 10, $k_1 \approx 0.338K$, $k_2 \approx 0.246K$ and $k_4 \approx 2.9105K$.

If we take into account only spin dipole excitations in the above Hamiltonian H, the number of quenching points in the interval h from 0 up 0.25 is three [2] (see Fig.3). According to experimental data, and by numerical diagonalization of the appropriate model Hamiltonian, only four quenching points have arisen [10].

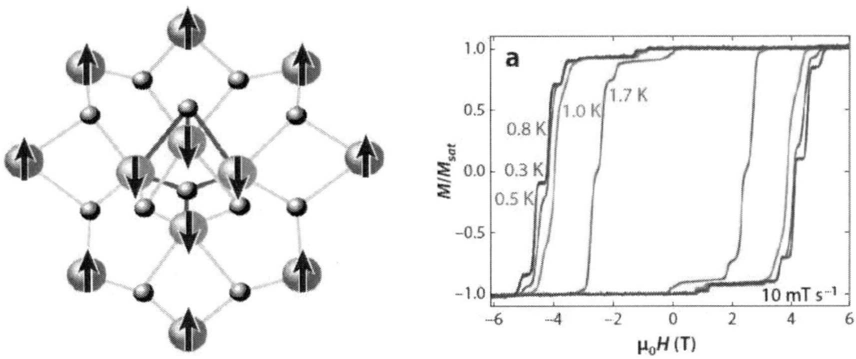

Figure 1. Nano particle Mn_{12} and Hysteresis Loop.

The effective model Hamiltonian of Mn_{12}-ac could be written as [10]

$$\hat{H}_{Mn} = \hat{J}^2 - \hat{J}_z^2 + k(\hat{J}_+^4 + \hat{J}_-^4) + \lambda(\hat{J}_x^2 - \hat{J}_y^2), \qquad (1.2)$$

where by $\hat{J}$ standard spin operators are denoted. We restrict the values of k and λ such that the z axis is the easy axis of spins. If we take into account only dipole excitations in the above Hamiltonian, the graphic in Fig.4 for spin tunneling is obtained.[3]

Figures 1 and 2 show typical hysteresis loops for both nano-particles, with the field applied along the easy axis of magnetization. Steps in the hysteresis

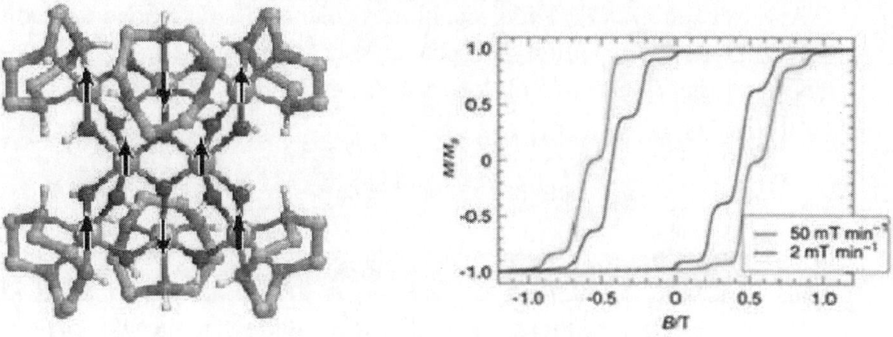

Figure 2. Nano particle Fe_8 and Hysteresis Loop.

loop are clearly visible. Up to date, it is suggested that the observed steps in hysteresis loops arise due to the resonance tunneling between quantum spin states in these nanoparticles and in calculations only dipole excitations have been considered [2,3]. The main idea of this paper is that for large spin values in the above Hamiltonians the excitations of multipole degrees of freedom of spin dynamics take place, which could be considered by an appropriate choice of trial functions.

In this paper, first we calculate the periodical dependence of spin tunneling for both nano-particles Fe_8 and Mn_{12} by use of the generalized coherent states constructed on SU(2) group as a basis of trial functions. In this case only dipolar degrees of freedom of spin dynamics could be taken into account. But due to the presence of single-ion anisotropy in Hamiltonians and large spin values here the excitation of multipole degrees of freedom of spin dynamics arises, which could be taken into account by use of the generalized coherent states constructed on SU(2S+1) group for the spin S [11]. In this paper we restrict ourselves by consideration of dipole and quadrupole excitations only. In order to investigate tunelling phenomenon between the ground states we use the instanton approach, which is describe in the next section.

2. Theory and Calculations

The instanton method is an efficient way of calculating tunnel splitting for spin systems. Instantons are classical paths that run among degenerate classical minima of the energy. By "classical" we mean that the path obeys the principle of

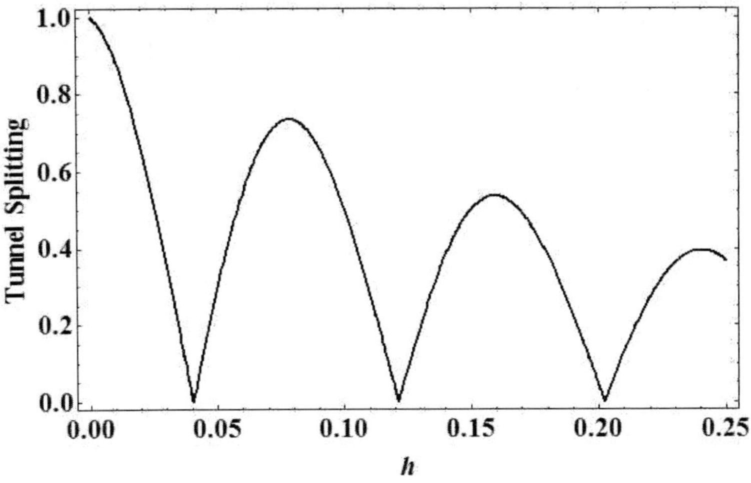

Figure 3. Tunnel splitting amplitude $\triangle$ versus h for Hamiltonian H_{Fe} in SU(2) group.

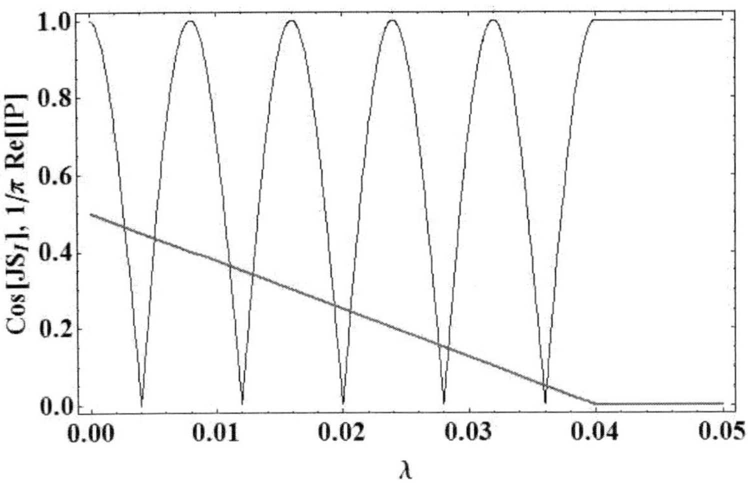

Figure 4. Tunnel splitting oscillation versus λ for Hamiltonian H_{Mn} in SU(2) group.

least action and satisfies energy conservation [1, 2].

When the action functional is specified, the instanton recipe for calculating the tunnel splitting is as follows. Let there be a number of instantons, i.e., least action paths, labelled by k, and let the actions for these various paths be S_k^{cl}. The tunneling amplitude is given by

$$\triangle = \sum_k D_k exp(-S_k^{cl}) \tag{2.1}$$

The prefactor D_k results from integrated the Gaussian fluctuations around the k-th instanton. For nano-particle Fe_8 , the tunneling amplitude is calculated [2,13] and the following relation is obtained:

$$\triangle \approx 2D_1 exp(-S_R) cos(J\Theta), \quad h < h_0 \tag{2.2}$$

In particular, there will be no quenching in $\triangle$ for $h_x > h_0$ where $h_0 = .25$. For $h_x < h_0$, quenching will occur when Berry phase is $\Theta = (2n+1)/2J$, where n is an integer number.

Applying SU(3) generalized coherent states technique to the Hamiltoniain (1.1) we obtain the classical energy for nano-particle Fe_8 is in the following form, where Euler parameters θ and ϕ are related to spin dipole excitation and parameter g is related to the quadrupole excitation [11]:

$$\begin{aligned} E_{Fe}(\theta,\phi,g) = & cos^2 2g cos^2\theta + \lambda_1 cos^2 2g sin^2\phi sin^2\theta - 2h cos2g cos\theta \\ & -2\lambda_2 cos^4 2g (cos^4\theta + sin^4\theta sin^4\phi - 6 sin^2\theta cos^2\theta sin^2\phi) \end{aligned} \tag{2.3}$$

By use of the approximate methods, we could obtain the minimum of energy. The instanton solutions could be found by exploiting energy conservation. Let us rewrite the condition $E(\theta,\phi,g) = E_{min}$ as follows

$$Z(g,\phi) cos^4\theta + R(g,\phi) cos^2\theta - 2h cos2g cos\theta + W(g,\phi) = 0, \tag{2.4}$$

where

$$\begin{aligned} Z(g,\phi) &= -2\lambda_2 cos^4 2g (1 + sin^4\phi + 2 sin^2\phi) \\ R(g,\phi) &= cos^2 2g (1 - \lambda_1 sin^2\phi) + 4lambda_2 cos^4 2g sin^2\phi (3 + sin^2\phi) \\ W(g,\phi) &= cos^2 2g sin^2\phi (\lambda_1 - 2\lambda_2 cos^2 2g sin^2\phi) + h^2 + 2\lambda_2 h^4. \end{aligned} \tag{2.5}$$

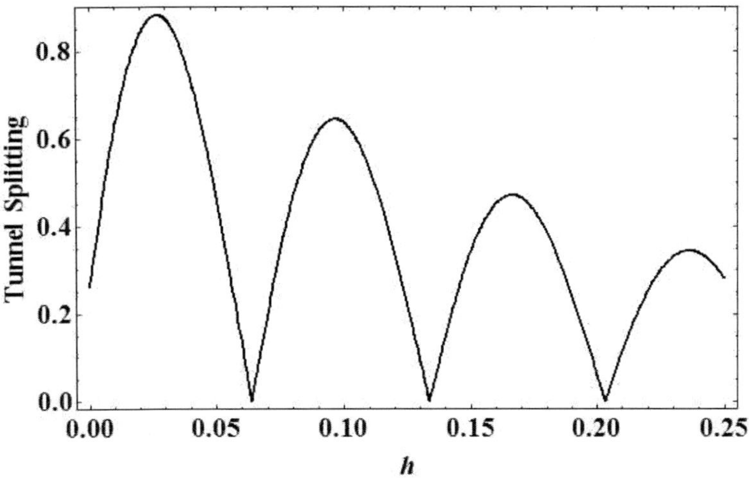

Figure 5. Tunnel splitting amplitude $\triangle$ versus h for Hamiltonian H_{Fe} in SU(3) group.

Then, let us consider the complex solution for the equation (2.4) in the form $cos\theta = A + iB$. Substitution of this solution in the equation (2.4) resulted in the following relation for real and imaginary parts of the action

$$\begin{aligned} S_R = ReS_B &= \int_0^{\pm\pi} Bd\phi \approx K(\lambda_1, \lambda_2) \\ \Theta &= \int_0^{\pi} (cos2g - A)d\phi - \int_0^{\pi} (1 - cos2g)d\gamma \\ &\approx \pi^2 - \int_0^{\pi}\int_0^{\pi} Adgd\phi \end{aligned} \tag{2.6}$$

By numerical calculation the following relation was obtained:

$$\triangle \approx 2D_1 exp(-4.5036h)cos(J(\pi^2 - 4.5036h)). \tag{2.7}$$

The function (2.7) is shown in the Fig.5. Comparing it with the Fig.3, one can conclude, that for this nano-particle taking into account of the quadrupole excitation shifts the location of the quenching points.

In the case of nano-particle Mn_{12}, the tunneling amplitude calculated in the frameworks of the SU(2) spin coherent states approach has the following form

[4]

$$\triangle \approx 4e^{-S_R}cos(J\pi)cos(J\Theta), \tag{2.8}$$

where only spin dipole excitations are taken into account. Applying the SU(3) generalized coherent states to the Hamiltonian (1.2) in addition to dipole excitations also quadrupole ones could be considered, and the classical energy takes the following form

$$\begin{aligned} E_{Mn}(\theta,\phi,g) = & J^2(1 - sin^2\theta cos^2 2g sin^2\phi) + \lambda J^2 cos^2 2g(cos^2\theta - sin^2\theta cos^2\phi) \\ & + 2kJ^4 cos^4 2g(cos^4\theta + sin^4\theta cos^4\phi - 6cos^2\theta sin^2\theta cos^2\phi) \end{aligned} \tag{2.9}$$

Carrying out the calculations similar to the previous section and making the substitution $P = Cos\theta$, we will seek the instanton solutions, which satisfy the constraint $E(\theta, \phi, g) = E_{min}$. Taking into account the boundary conditions we find the solution in the following form

$$P(\phi, g) = \sqrt{\frac{A(\phi,g) + i\sqrt{B(\phi,g)}}{C(\phi,g)}}, \tag{2.10}$$

where

$$\begin{aligned} A(\phi,g) = & 2acos^2\phi sec^2 2g(7 + cos2\phi) - \frac{1}{2}(1 + 3\lambda + (\lambda - 1)cos2\phi)sec^4 2g \\ B(\phi,g) = & 8a(1 + 6cos^2\phi + cos^4\phi)(1 + 2kcos^4 2g cos^4\phi - cos^2 2g(\lambda cos^2\phi \\ & + sin^2\phi)) - (\lambda + sin^2\phi + cos^2\phi(\lambda - 2acos^2 2g(7 + cos2\phi)))^2 \\ C(\phi,g) = & 4a(1 + 6cos^2\phi + cos^4\phi)cos^2 2g \end{aligned} \tag{2.11}$$

Substituting relations (2.11) in the equation (2.10) and using the minimum value of energy, one can calculate the splitting of the energy levels. In this section, we consider both dipole and quadrupole excitations of spin system. In order to conduct the numerical estimation of the splitting of energy levels we put $a = 10^{-4}$ and $k = 10^{-6}$. Like the previous section, the function $Cos(J\Theta)$ versus λ was plotted (see Fig. 6).

Comparing the number of quenching points for the nano-particle Mn_{12} in the received plot (Fig. 6) with the number of steps in the hysteresis loop (Fig. 1) one can see that these numbers coincide. Thus, accounting of the quadrupole dynamics excitations in the spin tunneling effect could give a deeper understanding of the essence of the phenomenon and better explanation of the experimental data.

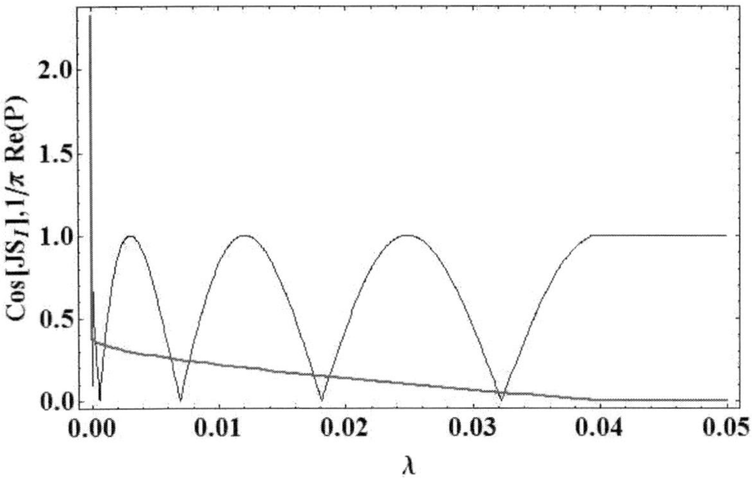

Figure 6. Tunnel splitting oscillation versus λ for Hamiltonian H_{Mn} in SU(3) group.

3. Conclusion

Thus in this paper we study the spin tunneling effects in nano-particles Fe_8 and Mn_{12} by instanton calculation technique using the SU(2) and SU(3) generalized spin coherent states as a trial function. It is established that in nano-particle Fe_8 taking into account of the quadrupole excitations shifts the location of the quenching points, but the number of these points depend on excitation of other degrees of freedom of multipole spin dynamics, which is out of the scope of this paper. In nano-particle Mn_{12}, as one can see from Figs.4 and 6, when the quadrupole excitations are taken into account, the number of quenching points reduces from 5 to 4. This number is equal to the steps in the experimental hysteresis loop of this crystal. Thus, the quadrupole excitations due to the presence of single-ion anisotropy in Hamiltonians and large spin value play a significant role in the semi-classical spin dynamics and spin tunneling phenomena in magnetic nanomolecules.

References

[1] S. P. Gubin, Yu A. Koksharov, G. B. Yurkov, *Russian Chemical Reviews*, 74(6), 489-520, 2005.

[2] Garg A., 1995, *Phys. Rev. B*, 51, 15161-69.

[3] Foss-Feig M. S. and Friedman J. R., 2009, *EPL*, 86, 27002.

[4] Friedman J. R., Sarachik M. P., Tejada J. and Ziolo R., 1996, *Phys. Rev. Lett.* 76, 3830-33.

[5] Barra A. L., Debrunner P., Gatteschi D., Schulz C. E., and Sessoli R., 1996, *Europhys. Lett.* 35, 133 38.

[6] Sangregorio C., Ohm T., Paulsen C., Sessoli R. and Gatteschi D., 1997, *Phys. Rev. Lett.* 78, 4645-48.

[7] Wieghardt K., Pohl K., Jibril I. and Huttner G. 1984 *Angew. Chem. Int. Ed. Engl.* 23, 77.

[8] Delfs C. et al. 1993, *Inorg. Chem.*, 32, 3099.

[9] Barra A. L., Debrunner P., Gatteschi D., Schultz CH. D. and Sessoli R. 1996 *Europhys. Lett.* 35, 133.

[10] Wernsdorfer W. and Sessoli R., 1999, *Science*, 284, 133.

[11] Abdulloev Kh. O. and Muminov Kh. Kh. 1994 *Phys. Solid state* 36.

[12] Langer J. S. 1967 *Ann. Phys.*, 41, 108.

[13] Garg A., 1993, *Europhys. Lett.*, 22, 205-10.

In: Computational Materials ... ISBN: 978-1-63482-541-2
Editor: K. T. Kholmurodov, pp. 63-71

Chapter 7

A SIMULATION STUDY ON CONFORMATIONS OF AN ANTI-MICROBIAL PEPTID

***Takao Yoda*[1,2,*] *and Toshiyuki Tsuji*[1]**
[1]Department of Computer Bioscience,
Nagahama Institute of Bio-Science and Technology,
Tamura, Nagahama, Shiga, Japan
[2]RIKEN Advanced Institute for Computational Science,
Chuo-ku, Kobe, Hyogo, Japan

Abstract

Conformations of an anti-microbial peptide cryptdin-4 and its disulfide-bridge mutants have been explored by molecular dynamics simulations. To ensure efficient samplings of conformations, we adopted replica-exchange method. The simulation results provided an insight into the mechanism of the anti-microbial activity of the disulfide-null mutant, which is known to be active in spite of its disordered nature.

Keywords: Molecular Dynamics Simulation, Replica-Exchange Method, Conformations, Anti-Microbial Peptide, Cryptdin-4, Disulfide-Bridge Mutants

1. Introduction

Molecular dynamics simulation is widely used for sampling of conformations of bio macromolecules like proteins and nucleic acids. Because many impor-

*E-mail address: t_yoda@nagahama-i-bio.ac.jp

tant physiological phenomena accompany large conformational changes of protein molecules, efficient and wide sampling of conformations of such macromolecules is considerably important to understand the mechanisms of such phenomena. To achieve an efficient and wide sampling of conformations, one can use generalized-ensemble algorithms. In a generalized-ensemble simulation, artificial weight factors are used instead of common ones (e.g. Boltzmann factor that is used to realize a canonical ensemble). Well-chosen artificial weight factors lead enhanced random walks in a parameter space and it eventually enhances the sampling of conformations by the simulation. Multicanonical algorithms (MUCA) [1] and replica-exchange method (REM) [2, 3] are ones of the most popular generalized-ensemble techniques. In MUCA and REM, random walks in the potential-energy space and in the temperature space are realized, respectively. Protein folding simulation is one of the most typical applications of such algorithms. We reported several successful applications of those methods and their derivatives, which include multicanonical replica-exchange method [4], for protein (or peptide) folding problems [5, 6, 7, 8]. It was shown that those simulation methods considerably reduce the simulation time required to observe folding events.

Here we describe our recent results of a replica-exchange simulation of an anti-microbial peptide (AMP). AMPs are polypeptides that have a physiological activity to kill microbes [9, 10]. Though their mechanisms of the anti-microbial activity are considered to be diverse, many of them perturb, or destroy, cell membranes of the target cells.

Some AMPs are known to remain active even when they are unfolded [11, 12]. We believe that we can utilize generalized-ensemble algorithms to understand the mechanism of activity of such disordered AMPs.

2. Materials and Methods

The molecule that we studied here is cryptdin-4 (Crp-4) [13]. It is an AMP of mice with 32 amino-acid residues. The primary sequence of Crp-4 is GLLCYCRKGCKRGERVRGTCGIRFLYCCPRR in one-letter code. We performed replica-exchange molecular dynamics (REMD) simulations of wild type (WT) and two mutant molecules of Crp-4. We used NAMD software [14] and CHARMM c36 force field [15] for the peptide molecule. We also used an implicit (generalized-Born) representation for water solvent that is available in NAMD software [16, 17]. The time step for integration was 2 fs. All bonds each of which connects a hydrogen atom to another atom were constrained by SHAKE. The number of replicas was 16 and the length of the production run was 60 ns per replica. The values of the 16 temperatures were 281.3, 300.0, 319.6, 340.1, 361.7, 384.2, 408.0, 432.9, 459.1, 486.6, 515.4, 545.7, 577.5, 610.9, 646.0, and 650.0 K. Exchanging temperatures between replicas was attempted every 100 fs. We analyzed conformations observed at 300 K.

We performed simulations in two different conditions. Only in one of the two conditions, a constant external electric field of -z direction was applied. This was done for mimicking the environment in the aqueous phase near a negatively charged membrane. The intensity of the external electric field applied (1.5×10^7 N/C) was determined as described below.

We assumed a system with continuum water solvent, ions, and a uniformly charged plane that is a model of a membrane. We did not put the peptide molecule here. To evaluate the intensity of the electric field near membrane based on this simplest model, one can use the following one-dimensional Poisson-Boltzmann equation [18, 19]:

$$\epsilon_0 \chi_{\mathrm{w}} \frac{\partial^2 \phi}{\partial z^2} = 2en_0 \sinh \frac{e\phi}{k_{\mathrm{B}}T}, \tag{2.1}$$

where $z, \phi, \epsilon_0, \chi_{\mathrm{w}}, e, n_0,$ and T are distance from the charged plane, electrostatic potential, vacuum permittivity, dielectric constant of water, elementary charge, bulk concentration of ions, and temperature, respectively. The solution of eq. (2.1) can be written as below [18, 19].

$$\phi(z) = -\frac{2k_{\mathrm{B}}T}{e} \ln \frac{1 + \gamma e^{-z/\lambda}}{1 - \gamma e^{-z/\lambda}}, \tag{2.2}$$

where

$$\gamma^2 + \frac{2b}{\lambda}\gamma - 1 = 0, \gamma > 0,$$

$$\lambda \equiv \sqrt{\frac{\epsilon_0 \chi_{\mathrm{w}} k_{\mathrm{B}} T}{2n_0 e^2}},$$

and

$$b \equiv \frac{2\epsilon_0 \chi_{\mathrm{w}} k_{\mathrm{B}} T}{e|\sigma|}.$$

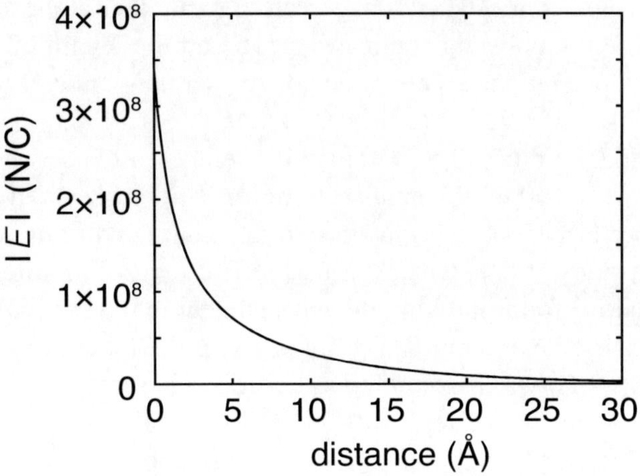

Figure 1. Electric field intensity calculated by eq. (2.2), plotted as a function of the distance from the uniformly charged plane.

Here, σ is the charge density on the plane that is placed at $z = 0$. According to the value of the electric field at $z = 15$ Å that was calculated assuming $\sigma = -0.0154e/\text{Å}^2$, $T = 300$ K, and $n_0 = 0.14$ M (see Fig. 1), we decided the intensity of the external electric field to be used in our simulation.

3. Results and Discussion

We show results of REMD simulations of WT and two mutants of Crp-4. One of the two mutants is the disulfide null mutant, in which all the Cys residues found in WT are substituted by Ala residues. The natively folded state of this mutant is known to be highly unstable because of the loss of disulfide bridges. The other mutant is C6/21A, in which Cys6 and Cys21 are substituted by Ala. It had been shown by experiments that the disulfide null mutant has considerable anti-microbial activity in spite of the disordered nature [12]. On the other hand, C6/21A has lower activity than both WT and the disulfide-null mutant [12].

3.1. Conformational Diversity of WT and Mutants of Crp-4

We estimated conformational diversities of WT and mutants Crp-4 by means of conformational clustering. We categorized conformations observed at 300 K in REMD trajectories by group average method. Before conformational clustering, conformation set obtained from the REMD under the external electric field was merged with that obtained from the REMD without the external electric field. Hierarchical clustering process was stopped when the minimum distance (C^{α}-RMSD) between clusters exceeded a threshold of 6.0 Å. Results are summarized in Table 1.

Table 1. Results of conformation clustering analyses.

	The number of conformation clusters †		
peptide	in EF *	w/o EF **	intersection ‡
WT	4	3	2
C6/21A	13	19	12
disulfide-null mutant	135	171	76

† The number of clusters when hierarchical clustering was stopped with the threshold distance of 6.0 Å between clusters.

* The number of conformation clusters that contain conformation(s) taken from the REMD in the presence of the external electric field.

** The number of conformation clusters that contain conformation(s) taken from the REMD in the absence of the external electric field.

‡ The number of conformation clusters that contain both conformation(s) from the REMD in the presence of the external electric field and conformation(s) from the REMD in the absence of it.

The results indicate that the removal of disulfide bridge(s) increases flexibility and diversity of the peptide conformation, as expected. When we compare the two conditions for each peptide, a lot of conformers observed in the absence of the external electric field turned out not to be observed in the presence of the external electric field. This suggests that the external electric field largely altered the distribution of conformations of the disulfide-null mutant. Thus in order to understand mechanisms of these peptides functions, it is meaningful to investigate conformations observed under such a condition with the external electric field that mimics the environment near a membrane.

3.2. Hydrophobic Cluster Formation

AMPs typically have an amphiphilic nature to interact with lipid membrane. Thus hydrophobic residues on the surface of AMPs play important roles in their functions. To understand effects of the large flexibility due to the removal of disulfide bridges, we evaluated spatial distribution of hydrophobic side chains. The results show that hydrophobic side chains in the molecule of the disulfide-null mutant highly localize on one side that is opposite to the direction of the external electric field (see Fig. 2). On the other hand, in WT and C6/21A this localization of hydrophobic side chains are not so evident as the disulfide-null mutant.

These results suggest that the enhanced flexibility of the molecule of the disulfide-null mutant helps the hydrophobic side chains localization at least in the presence of the external electric field. In addition, comparison between the disulfide-null and the C6/21A mutants implies that the removal of a disulfide bridge between Cys6 and Cys21 is not enough for a hydrophobic side chains localization.

3.3. Implication on Activity of Mutants

In the present simulation results, changes in the conformation distribution due to the electric field were observed, especially in the disulfide-null mutant. Spatial distribution of hydrophobic side chains turned out to depend remarkably on the molecule type. Enhanced localization of hydrophobic side chains in the disulfide-null mutant can be due to its conformational flexibility. Increased localization of hydrophobic side chains should increase the interaction of the peptide with membrane. This may be a cause of the remaining activity of the disulfide-null mutant of Crp-4 [12].

The present findings imply that the removal of disulfide bridges brought about a kind of nonnative conformational order, or residual structure, in some portion(s) of the flexible molecule.

4. Conclusion

We have reported our preliminary results of simulations of the wild type and two mutants of an antimicrobial peptide cryptdin-4. We used replica-exchange method for efficient conformational sampling. To mimic the environment in

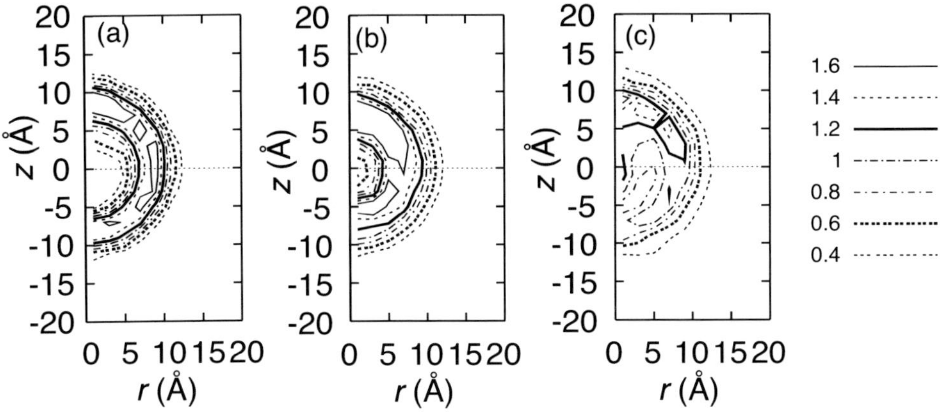

Figure 2. Two-dimensional contour maps of number density of hydrophobic side chains of Crp-4 and its mutants in the external electric field applied. Results for WT (a), C6/21A (b), and the disulfide-null mutant (c) are shown. The unit of the number density is nm^{-3}. Before the evaluation of the number density, the peptide molecule was translated so that the center of mass of the molecule comes to the origin of coordinates. The horizontal axis represents the distance from the z axis (r).

aqueous phase near a charged membrane, constant external electric field was applied.

The change in the conformation distribution in response to the external electric field is likely due to interaction of positively charged amino-acid residues with the external electric field. Thus further analyses are necessary to illustrate more clearly the relationship between the mutations and the antimicrobial activities. They will be reported elsewhere [20].

Acknowledgments

This work was supported by the Platform for Drug Discovery, Informatics, and Structural Life Science from the Ministry of Education, Culture, Sports, Science and Technology, Japan (MEXT), and JSPS KAKENHI Grant Number 25330355.

References

[1] Berg, B.A., Neuhaus, T., Multicanonical ensemble: a new approach to simulate first-order phase transitions. *Phys. Rev. Lett.* 68, 9-12 (1992).

[2] Hukushima, K., Nemoto, K., Exchange Monte Carlo method and application to spin glass simulations.*J. Phys. Soc. Jpn.* 65, 1604-1608 (1996).

[3] Sugita, Y., Okamoto, Y., Replica-exchange molecular dynamics method for protein folding.*Chem. Phys. Lett.* 314, 141-151 (1999).

[4] Sugita, Y., Okamoto, Y., Replica-exchange multicanonical algorithm and multicanonical replica-exchange method for simulating systems with rough energy landscape.*Chem. Phys. Lett.* 329, 261-270 (2000).

[5] Sugita, Y., Okamoto, Y., Molecular mechanism for stabilizing a short helical peptide studied by generalized-ensemble simulations with explicit water.*Biophys. J.* 88, 3180-3190 (2005).

[6] Yoda, T., Sugita, Y., Okamoto, Y., Cooperative folding mechanism of a β-hairpin peptide studied by a multicanonical replica-exchange molecular dynamics simulation.*Proteins* 66, 846-859 (2007).

[7] Yoda, T., Sugita, Y., Okamoto, Y., Hydrophobic core formation and dehydration in protein folding studied by generalized-ensemble simulations.*Biophys. J.* 99, 1637-1644 (2010).

[8] Yoda, T., Sugita, Y., Okamoto, Y., Salt effects on hydrophobic-core formation in folding of a helical miniprotein studied by molecular dynamics simulations.*Proteins* 82, 933-943 (2014).

[9] Brogden, K.A., Antimicrobial peptides: Pore formers or metabolic inhibitors in bacteria?*Nat. Rev. Microbiol.* 3, 238-250 (2005).

[10] Ouellette, A.J., Paneth cell α-defensins in enteric innate immunity.*Cell. Mol. Life. Sci.* 68, 2215-2229 (2011).

[11] Wu, Z., Hoover, D.M., Yang, D., Boulègue, C., Santamaria, F., Oppenhaim, J.J., Lubkowski, J., Lu, W., Engineering disulfide bridges to dissect antimicrobial and chemotactic activities of human β-defensin 3.*Proc. Natl. Acad. Sci. USA* 100, 8880-8885 (2003).

[12] Maemoto, A., Qu, X., Rosengren, K.J., Tanabe, H., Henshen-Edman, A., Craik, D.J., Ouellette, A.J., Functional analysis of the α-defensin disulfide array in mouse cryptdin-4.*J. Biol. Chem.* 279, 44188-44196 (2004).

[13] Ouellette, A.J., Hsieh, M.M., Nosek, M.T., Cano-Gauci, D.F., Huttner, K.M., Buick, R.N., Selsted, M.E., Mouse paneth cell defensins: Primary structures and antibacterial activities of numerous cryptdin isoforms.*Infect. Immun.* 62, 5040-5047 (1994).

[14] Phillips, J.C., Braun, R., Wang, W., Gumbart, J., Tajkhorshid, E., Villa, E., Chipot, C., Skeel, R.D., Kale, L., Schulten, K., Scalable molecular dynamics with NAMD.*J. Comput. Chem.* 26, 1781-1802 (2005).

[15] Best, R.B., Zhu, X., Shim, J., Lopes, P.E.M., Mittal, J., Feig, M., MacKerell, A.D., Optimization of the additive CHARMM all-atom protein force field targeting improved sampling of the backbone ϕ, ψ and side-chain χ_1 and χ_2 dihedral angles.*J. Chem. Theory Comput.* 8, 3257-3273 (2012).

[16] Onufriev, A., Bashford, D., Case, D.A., Modification of the generalised born model suitable for macromolecules.*J. Phys. Chem.* 104, 3712-3720 (2000).

[17] Onufriev, A., Bashford, D., Case, D.A., Exploring protein native states and large-scale conformational changes with a modified generalized born model.*Proteins* 55, 383-394 (2004).

[18] Andelman, D., Electrostatic properties of membranes: The Poisson-Boltzmann theory. In *Handbook of biological physics*, 1 , pp. 603-642 (1995).

[19] Israelachvili, J.N., Intermolecular and surface forces, Academic Press, London, (1990).

[20] Yoda, T. *et al.*, manuscript in preparation.

In: Computational Materials ...
Editor: K. T. Kholmurodov, pp. 73-81
ISBN: 978-1-63482-541-2

Chapter 8

MATHEMATICAL MODELING OF THE EFFICIENCY GADOLINIUM BASED NEUTRON CONVERTERS

Djamshed A. Abdushukurov*
Physical-Technical Institute named after S.U. Umarov
Academy of Sciences of Republic of Tajikistan
Dushanbe, Tajikistan

Abstract

Procedures of mathematical modeling of thin solid-state converters of neutrons are described. As converters a foil from natural gadolinium and its 157 isotopes are viewed. Algorithms of calculation of absorption of neutrons in a material of converters, probabilities formation of secondary electrons, probability of an exit of electrons from a material of the converter and the efficiency of converters are given. Calculation was made for various thicknesses of converters, and various wavelengths of neutrons. The optimal converter thicknesses were chosen.

Keywords: Mathematic modeling, Neutron converter, Gadolinium

---

* E-mail address: abdush_dj@mail.ru

1. Introduction

For detection of neutrally charged neutrons, it is necessary to use converters of the neutron radiation. These converters are converse radiations of neutrons to the charged radiation, which can be in a further detected with various detectors. For an optimal choice of the converter foil material, the following properties should be consider:

- A large neutron capture-cross section needed in order to achieve high detection efficiency with minimum foil thickness.
- The range of the neutron-induced charged primary particles should be large compared with the convertor foil thickness.
- The converter material should possess small cross-section of activation. Daughter isotopes occurring as result of activation should not create the significant background.
- The γ-ray sensitivity of the detector material should be low.

Mathematical modeling allows to prospect, without special material cost, to carry out searching the most suitable configuration of converters and the optimum geometrical ratio. It in turn allows refusing from realization of superfluous development and researches.

2. Model Calculation. Results and Discussion

Efficiency of the Gadolinium based neutron converters is the ratio of the electrons which have depart from converters, to total number of falling on the converter neutrons. The efficiency of converters is the product of three variables. These are the following:

- Probabilities of thermal neutrons capture by nucleus of the converter, which depends on its thickness, and the cross-section of interaction. The cross-section of interaction in turn depends on isotope composition of the converter and energy or wavelength of neutrons.
- Probabilities of creation of the secondary charged particles, in our case of internal conversion and Auger electrons.

- Probabilities of exit of the created electrons from the substance of the converter, which depends on, free path length of electrons in a converter substance and thickness of converter.

Energies of the secondary electrons are discrete values and are formed with the certain probability. At calculation of probability of exit it is necessary to take into account emission of each electron with its characteristic energy and its weight factor (probability of formation). In that specific case efficiency can be described by the following expression [1,2]

$$Wi = Pn(\sigma Lj) * De(EiNi) * He(ReiLi) \tag{2.1}$$

Where Wi is the efficiency of registration of the absorbed with the probability $Pn(\sigma Lj)$ neutrons, which caused formation of electrons $De(EiNi)$ with the discrete energies and probability of it‘s formation, probability of escape of electrons from a material of converter $He(ReiLi)$ in view of the maximal free path length of electrons in a sub-stance of the converter and its thickness. Generally all probabilities are necessary for summarizing, so calculations are made for 446 discrete energies of electrons with the energies in the range from 0.9 up to $1000keV$.

The calculations was made for thermal neutrons with fixed energies, which correspond to the neutron wavelengths of 1, 1.8, 3 and $4\AA$; also both the thickness of converters from $1\ m\mu$ to $40\ m\mu$, and for the isotopic composition of converter (for natural Gd and ^{157}Gd) were varied.

Conventionally we were dividing the thickness of a foil into more thin components. For each elementary layer we count the probability of neutrons absorption with the fixed energy. Efficiency of the converter will be determined by the multiplication of probabilities of neutron absorption and probability of the electron exit from the converter. The choice of the model was made from the following assumptions: all electron emissions are isotropic, the length of the free path for any fixed energy of electrons (Rei) is constant. Then the density of probability to find electrons in the material forms is a sphere with radius equal to Rei (for each fixed energy). If the center of this sphere is crossed by the plane, two identical hemispheres are formed, which correspond to the electrons escape to the forward hemisphere and backward hemisphere, thus the volume of hemispheres could be considered as the probability of the electrons output. In this case the total probability is 100%, and exit to the one of the hemispheres is 50%. If we begin to cross a sphere with a step much less than Rei, segments will be

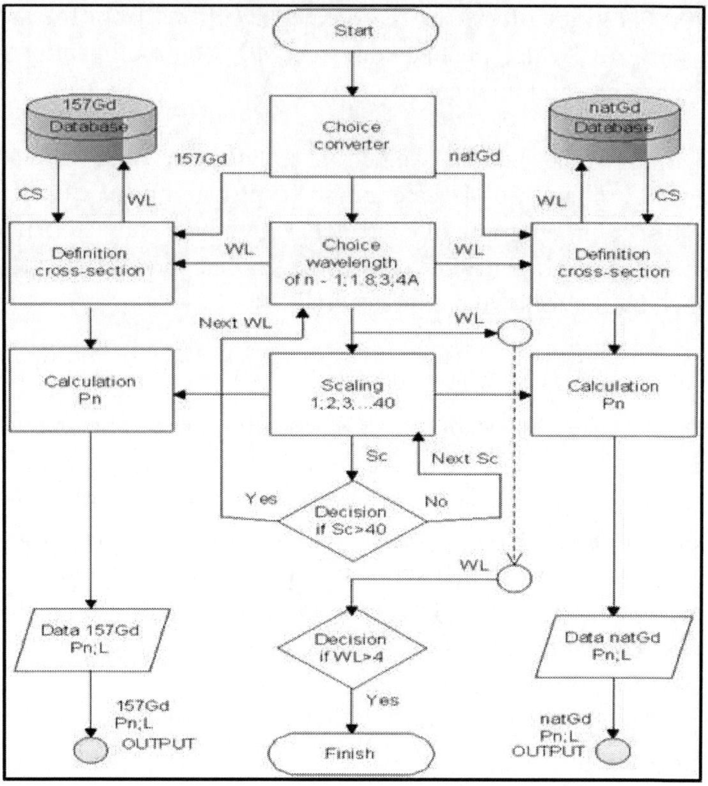

Figure 1. Calculation algorithm diagram of absorption path of neutrons in substance.

formed whose volume will be equal to the probability of the electron presence. The step of iterations should be less than Rei. The volume of a segment and accordingly probability of the electron escape becomes equal to zero in the intersection of a sphere by the plane at the distance Rei. The sum of probabilities of electron escapes for all energies, taking into account their weight contribution and will determine the total electron escape probability.

The scheme of the algorithms calculation of neutron absorption in a converter material is shown in Figure 1. In calculations, of the probability of neutron absorption in each elementary layer, as a result of reaction of inelastic interaction (n, γ), it is necessary using the database on cross-section of neutron interaction for each fixed energies. In the reaction of ^{157}Gd neutron capture,

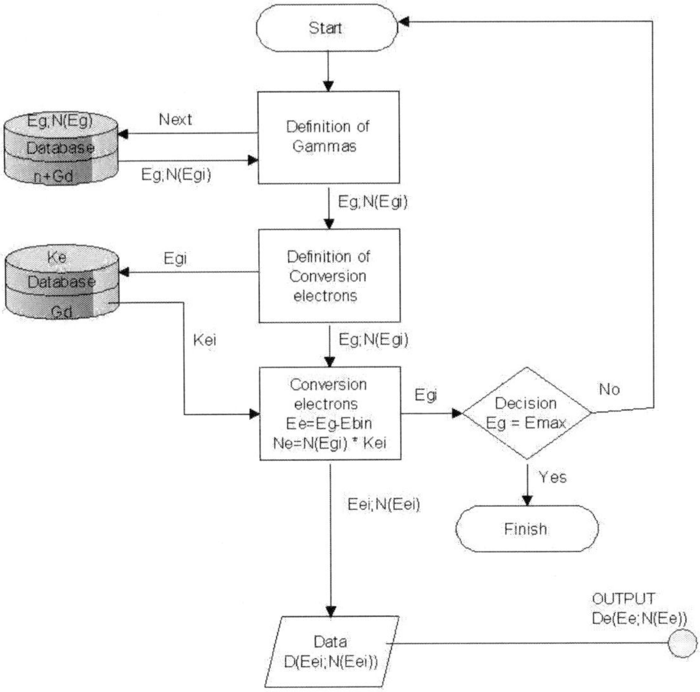

Figure 2. Diagram of algorithms for calculation of number of internal conversion electrons.

$7937.33keV$ energy is emitted. In total 390 gamma lines with energy in ranges from 79.5 up to $7857.670keV$ with line intensity of $2-10^{-8}$ up to 139 gamma-quanta on 100 captured neutrons are emitted.

The low-energy gamma quanta are present at the spectrum and during their emission electrons from atomic subshells (so-called internal conversion electrons) are radiated with a high probability. The nuclear removes from its excitation levels by radiating a gamma-quantum, but also a close located electron can be irradiated too. Usually K-electron (electron from K-subshell) is emitted, but also electrons from the higher subshells (like L, M, N and so on) can be emitted too. Vacancy of electrons (an electronic hole), formed as a result of this process, is filled by another electron from a higher levels. This process is accompanied by radiation of X-ray quantum, or radiation of $Auger$-electron. The probability of exit of electrons from a converter substance can be considered on

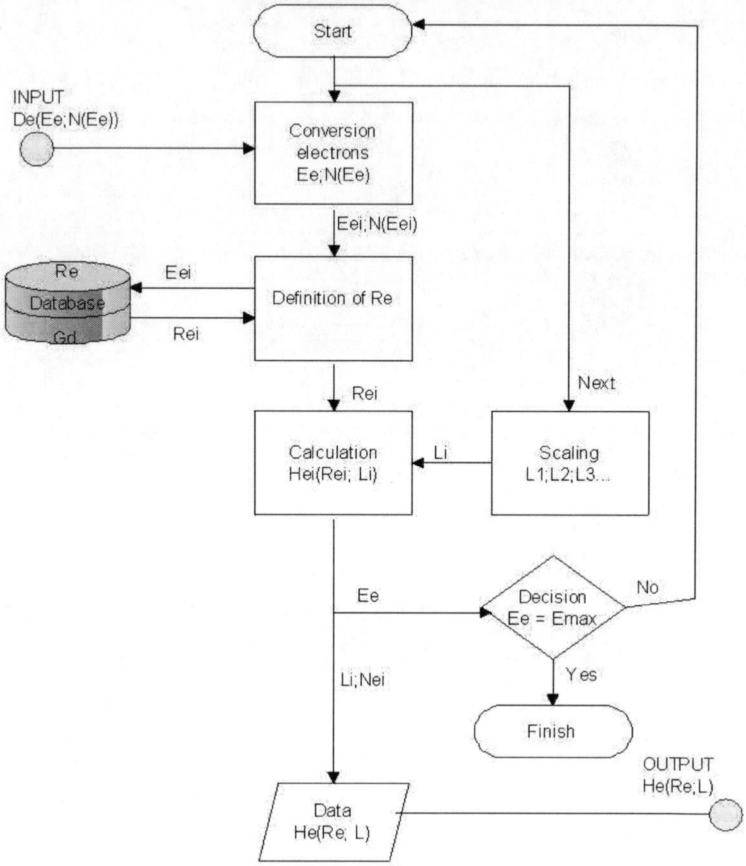

Figure 3. Scheme of algorithm calculation of ability an electrons exit from the converter body.

the basis of the Bethe-Bloch theory. For realization of calculations it is important to have exact data on brake ability for various substances for the charged particles. Brake ability is average speed of the loss of energy by charged particles in any points along their tracks. The scheme of algorithm calculation of ability an electrons exit from the converter is shown in figure 3. For calculations of probabilities of electrons output from substance the data about internal conversion and *Auger* downloaded. For each discrete energy of an electron the database about the free length (Re) electron in substance were used. For each elemental stratum probability of an exit of electrons is yielded, for calculation the data on Re and intensity of electrons used.

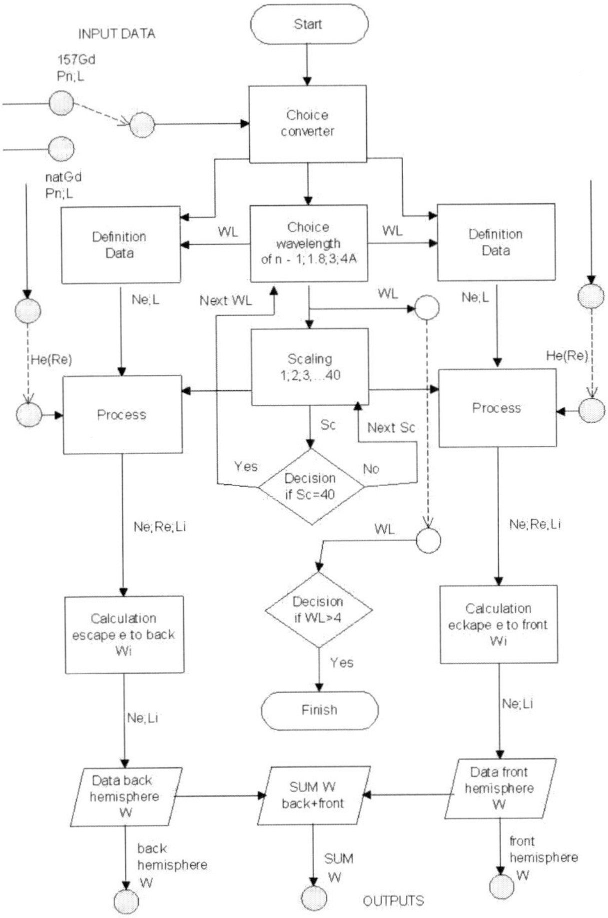

Figure 4. Scheme algorithm calculation of the efficiency of converters.

Calculations of converters efficiency for geometry of an exit of electrons in a back hemisphere, forward hemisphere and their sum for different thickness of the converter have been performed. The scheme of algorithms of calculation is given in figure 4. In the beginning the select of the type of converter and a neutron wavelength was made. The thickness of the converter is divided into thin elemental stratums. For each stratum the corresponding data on absorption of neutrons (Pn, L) is download. During calculation to each stratum conventionally co-ordinates $X0$, $Y0$ appropriated, the data of probability of an exit of electrons ($He(Re)$) is download. The data gained for each stratum are tots. During the modeling the calculations efficiency separately for geometry of an exit of electrons in a forward hemisphere and back hemisphere and their sum were made. At 30 microns thickness of the natural Gadolinium converter the neutrons with the wave length above $1.8\AA$ are absorbed completely. For its 157 isotope for neutrons with the wave lengths ¿$1.8\AA$ the same absorption happens at 8 microns thickness of the converter. In order to detect thermal neutrons by the ^{157}Gd converters we could limit by the thickness of 5 microns of the converter, if there no technological restrictions. It should be taken into account that the majority of electrons emitted in the reaction of radiating capture of neutrons have running less than 5 microns. When using natural Gadolinium the situation is more complex, since low absorption requires converter to be thicker than $20 - 40$ micron.

3. Conclusion

Mathematical modeling of efficiency of registration of thermal neutrons by the foil converters made from natural $Gadolinium$ and its 157 isotope carried out. Algorithms of calculations of absorption of neutrons in substance of the converter, probability for formation of secondary electrons and probability for an exit of electrons from the substance of converter are described. Processes of neutron absorption in the material of a converter and the probability of secondary electron escapes were examined. Calculation made for converters with the various thicknesses. We have chosen the most optimal converter thicknesses, both from natural Gadolinium and from its 157 isotope. While using converters from natural Gadolinium it is possible to obtain total efficiency of 10, 21, 26, 30%, correspondingly for the neutrons with wavelengths of 1, 1.8, 3 and $4\AA$ with converter thickness of 24, 7, 5, 4 microns. For the converters with 157 isotopes it is possible to reach total efficiency of registration up to 27, 45, 49, 52% for the neutrons with the wavelengths 1, 1.8, 3 and $4\AA$ with the thickness of converter 5, 3, 2 microns, respectively. The received results are in a good agreement with the experimental data.

References

[1] D.A. Abdushukurov, *Gadolinium Foils as Converters of Thermal Neutrons in Detectors of Nuclear Radiation*, ISBN: 978-1-61728-676-6, New-York, Nova Science Publishers Inc., 2010, pp. 1-124.

[2] D.A. Abdushukurov, Application of the Gadolinium Foils as Converters of Thermal Neutrons in Detectors of Nuclear Radiation. In: Caden C Tompson editor, *Gadolinium: Compounds, Production and Application*, Chapter 2, ISBN: 978-1-61668-991-9, New-York, Nova Science Publishers Inc., 2010, pp. 53-141.

[3] D.A. Abdushukurov, A.A. Djuraev, S.S. Evteeva et al. Model Calculation of Gadolinium- Based Converters of Thermal Neutrons, *Nucl. Instr. And Meth.*, 1994, v.B 84, p.400.

In: Computational Materials ... ISBN: 978-1-63482-541-2
Editor: K. T. Kholmurodov, pp. 83-96

Chapter 9

Atomistic Basis for Unique Physical Properties of Archaeal – Like Membranes (Discovered by Computer Simulations)

Anton O. Chugunov*, Pavel E. Volynsky, Nicolay A. Krylov, Ivan A. Boldyrev and Roman G. Efremov
M. M. Shemyakin and Yu. A. Ovchinnikov,
Institute of Bioorganic Chemistry, Russian Academy of Sciences,
Moscow, Russia

Abstract

Archaea differ from Bacteria and Eucaryota in chemical structure of plasma membrane lipids by the two most important features: (1) stereochemistry of glycerol backbone and (2) the structure of hydrophobic core chains. While former unlikely influences membrane properties, the latter (isoprenoid nature of the tails) was hypothesized to render archaeal membranes extremely durable and almost impermeable for water and ions. Additionally, these membranes do not exhibit apparent phase transition at physiological conditions and may be considered liquid-crystalline in $0 - 100 +^{\circ}$ C temperature range. We use molecular dynamics (MD) calculations to clarify the role of branching of these lipids hydrophobic

* E-mail address: batch2k@yandex.ru

tails (specifically, presence of side methyl groups and cyclopentane moieties) in membrane physical properties. From analysis of a series of archaeal tetraether lipids mimetics we conclude that the branched structure defines dense packing and low permeability, but at the same time liquid-crystalline state, which is vital for every living cell.

Keywords: Archaeal Biomembranes, Thermophiles, Phase Transition

1. Introduction

Archaea are microorganisms that thrive predominantly extreme conditions (very high/low temperatures, high pressure or salinity), although many of them inhabit more common econiches (ocean water, soil, intestine, etc) [1–3]. According to pioneering works by Carl Woese, Archaea are not only ecologically, but also evolutionary self-standing group, which nowdays is designated as third domain of life (along with Bacteria and Eucaryota). A hypothesis has been thrown that separation of Bacteria and Archaea resulted from individualization of membranes, which were promiscuous in Last Universal Common Ancestor (LUCA) and became more specialized in two modern domains of microorganisms due to so-called Lipid Divide ancient event [4]. There are two major differences in membrane lipids structure between Archaea and Bacteria: 1) glycerol backbone chirality and the way of connection to hydrophobic tails, and 2) chemical nature of these tails. In Bacteria and Eucaryota, straight fatty acid acyl chains are linked by ester bonds to sn-1 and sn-2 positions of glycerol, while in Archaea branched isoprenoid hydrocarbon chains are bound by ether bonds to sn-2 and sn-3 glycerol positions [5]. Glycerol backbones are enantiomers in Bacteria and Archaea. Polar heads used are also often different, although they are not unique to any domain of life.

Of two aforementioned chemical differences between bacterial and archaeal lipids, glycerol chirality may have played very important evolutionary role, while most probably it does not affect physical properties of the membranes. On the contrary, primordially branched isoprenoid chains in Archaea are likely to determine very dense packing, high membrane viscosity [9] and durability [10], and low permeability for water and ions, which are the main characteristics of archaeal membranes as compared to bacterial or eukaryotic ones [11–13]. Additional modification that increases density of packing and decreases membrane permeability are cyclopentane rings in hydrophobic tails, which are encountered more often in acidophilic microorganisms. Moreover, Archaea are able to adapt their membranes to better fit their growth conditions, employing several types of lipids structure variation. Finally, most hyperthermophiles possess lengthy C40-based glycerol-dialkyl-glyceroltetraether (GDGT) lipids, which are usually referred to as caldarchaeols [16] and represent tail-to-tail linked glycerol-dialkyl-glyceroldiether (GDGD) lipids, or archaeols. This modification provides extra durability for the membranes by switching from bilayer to more rigid monolayer structure.

In spite these extraordinary mechanical properties that remind of gel state of conventional dipalmitoylphosphatidylcholine (DPPC) [8], archaeal membranes may be considered liquid-crystalline in a very broad temperature range ($0 - 100 + ^\circ C$), which is believed to be vital for each living cell. Very high durability and liquidity at the same time seems to be the most fundamental physical property of archaeal membranes, and it not only provides insight into evolutionary fate of Archaea, but also may be adopted from the nature by bionanotechnology as a new unique material [19]. Rational design of such materials requires detailed understanding of the role of certain chemical groups in these properties. Although this task is within the reach of modern computational methods, there are few research papers published in this field. The most important work is by Bulacu et al. [20], who performed molecular dynamics (MD) simulation of bolalipid membrane in a coarse-grained representation. Due to this, the role of branching methyl and cyclopentane groups could hardly be assessed, although authors provide some insight into phase condition and design of synthetic durable membranes. In their early work Gabriel & Chong [21] assess by MD the role of cyclopentane rings in tight packing of the membrane, although they do not systematically address the issue of branched methyl groups. Another work has shown effect of tail linkage in archaeal lipids: expectedly, tetraether lipids exhibit lower lateral diffusion as compared to diether ones [22].

2. Results and Discussion

In this article, we describe a computational approach to rationalize the role of branching groups in formation of the mentioned unique properties of the archaeal membranes. To do this, we simulated several model GDGT monolayer membranes where the structure of hydrophobic tails gradually varied from straight C32-fatty acid acyl chains to archaeal-like C40 methyl-branched isoprenoid chains. In all cases tails were O-linked. Apart from methyl groups, cyclopentane moieties were also considered in calculation. Glycerol back-

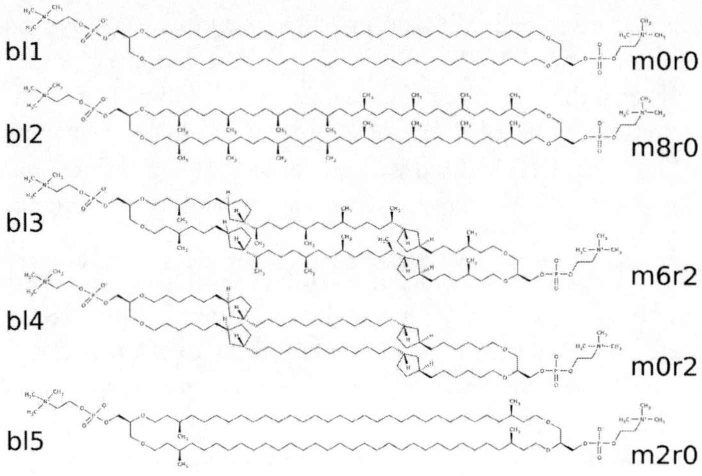

Figure 1. Chemical modifications of bolalipid mimetics hydrophobic tails that were used in MD simulations. The core structure represented with bl-1 consists of two straight O-linked C32-fatty acid acyl chains, connected to archaeal-type glycerol molecule with ether bonds. Hydrophilic heads are chosen to be phosphatidylcholine. Other mimetics (bl-2 – bl-5) possess different number of branching modifications of the hydrophobic core: methyl groups (designated as m in mimetic codes to the right) and cyclopentane rings (designated as r in mimetic codes).

bone stereochemistry and the way of connection to the hydrophobic tails were archaeal-like, whereas polar heads were chosen to be just phosphatidylcholine (the most common variant in Eucaryota), to bring in focus effect of chemical structure of hydrophobic chains. The five studied mimetics were symmetrical and modeled in a full-atom regime using GROMACS suite at four different temperatures (290, 310, 330 and 350 K) each. If not explicitly mentioned, all results hereinafter correspond to 310 K. Length of the each trajectory was $0.5\mu s$ with total $>10\mu s$ of MD statistics collected for this work. Molecules were arranged to form a 12x12 lipid monolayer (in total, 144 lipid molecules per model membrane; see Figs.1 and 2 for mimetics structure and Computational methods for simulation details).

The most important result of these simulations is seen by naked eye: completely straight fatty-acid hydrophobic tails in bl-1 membrane gain almost crystallographic ordering and strongly remind of gel phase, as compared to fully

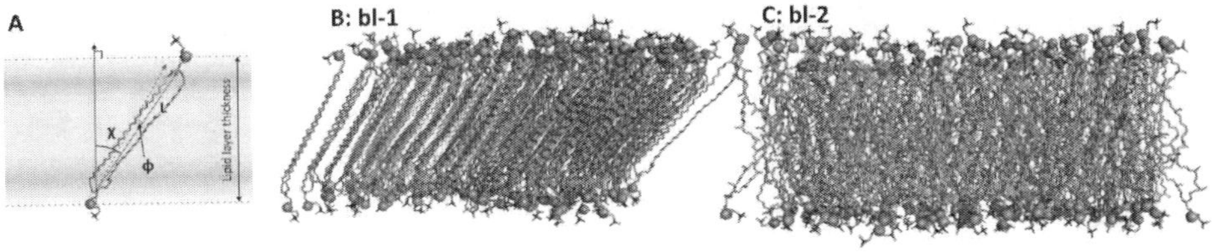

Figure 2. Model bolalipid membranes. A. Position of individual bolalipid molecule with respect to the model membrane (schematically depicted with yellow layer). Phosphorus atoms are shown with orange spheres; they define the thickness of the membrane. For analysis, each lipid was represented as a vector (blue arrow), built with two sn-2 atoms from opposite glycerol backbones. The length of this vector (L) describes conformation of the lipid (extended/expanded). Lipid tilt angle (χ) was measured with respect to layer normal (black arrow). Lipid-lipid collinearity angle (φ) was measured between each pair of vectors within one MD snapshot (the second vector is shown with blue broken arrow). B.Model membrane of bl-1 bolalipid mimetic (with completely straight tails). C. Model membrane of bl-2 mimetic, which contain branched methyl group in each fourth position of the hydrophobic tail. Note considerable ordering and tilting of lipids in bl-1 membrane as compared to bl-2.

branched bl-2 membrane (Fig.2B,C). Lipid molecules in bl-1 appear to be extended, tilted with respect to membrane normal and highly collinear with respect to each other. To characterize this quantitatively, we introduce parameters of lipid length (L), tilt angle (χ) and collinearity angle (φ), which are explained in Fig.2A and in Computational methods. We study MD-distributions of these parameters (Fig.3) to bring some formalism to the first sight analysis.

Analysis of Fig.3 provides evidence that branching of bolalipid tails plays an important role in phase behavior of model membranes, reminiscent of unsaturated bonds in bacterial and eukaryotic phospholipids. Membranes made of bl-2 and bl-3 mimetics that possess eight branching modifications in each tail (these are eight methyl groups and six methyl groups plus two rings, respectively; see Fig.1) form a liquid-crystalline phase. At the same time, bl-1, bl-4 and bl-5, which either do not have such modifications, or possess only two (two methyl groups and two rings for bl-4 and bl-5, respectively), form a gel-like phase. This phase is characterized with stretched hydrophobic tails (Fig.2B,C; Fig.3B), high lipid tilt angle with respect to the membrane normal (Fig.3A), almost complete collinearity of the molecules within MD box (Fig. 3C) and high-density packing of the hydrophobic core (Fig.3D). These results stay true for all temperatures studied.

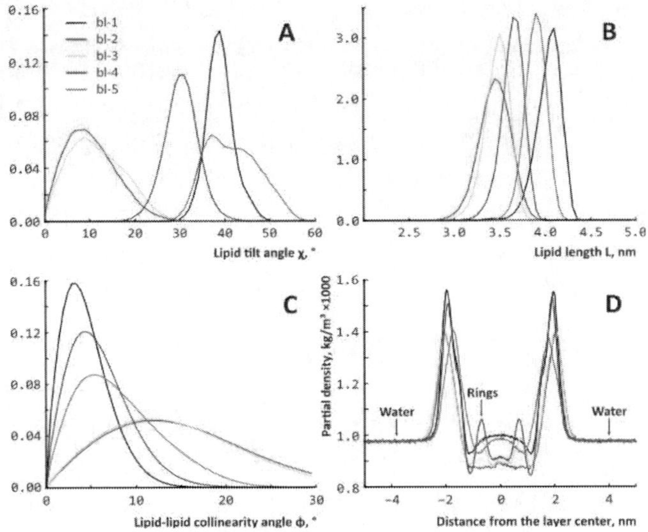

Figure 3. Packing parameters in model bolalipid membranes. Each bolalipid mimetic is shown with individual color, according to the legend. Panels A–C are probability density functions. All parameters are given for 310 K MD runs. A. Lipid tilt angle with respect to the layer normal (χ). Note that bl-2 and bl-3 are less tilted. B. Lipid length (L) which describes conformation and membrane parameters. Note that bl-2 and bl-3 have minimal, while bl-1maximal length. C.Lipid-lipid collinearity angle (φ) between all molecule pairs within one MD snapshot. Note that molecules in bl-2 and bl-3 are less collinear and more independent from each other. D. Partial density profiles of the whole systems (including solvent and lipid molecules) along the membrane normal. Note that in the center the least densely packed are bl-2 and bl-3, while bl-1 is the densest. The bl-3 and bl-4 profiles possess peaks from cyclopentane rings (shown with an arrow).

One of the hallmarks of archaeal membranes is the lack of distinct phase transitions and relatively weak dependence of properties on temperature. We have performed MD calculations at four different temperatures, and observe only slight alterations of computed properties. For example, bl-1 represents a gel-like phase at all temperatures, and bl-2/bl-3 are in a liquid-crystalline state in the whole temperature range studied. This feature may be an important prerequisite for Archaea to thrive in a wide range of econiches, from cold ocean water to high-temperature and pressure hydrothermal vents.

It has been shown that Archaea species that reside in low-pH and high-salinity conditions have more cyclopentane rings in their lipids tails, and this number may vary depending on the growth condition. In fact, maximum ring number is achieved at optimal for this species conditions. The main hypothesis is that cyclopentane rings additionally decrease permeability of the membrane for water and ions, which is vital for survival in such an extreme conditions. From partial density profiles in Fig.3D one can note sharp peaks which correspond to the cyclopentane rings, and this may further improve packing and decrease permeability.

Long-term MD trajectories permit assessment of water permeability of model membranes (see Computational Methods). Figure 4 shows the frequency of water permeation events (due to free diffusion) across the five model bolalipid membranes, and also across two common phospholipid bilayers for the purpose of comparison. A gel-like bl-1 membrane was completely impermeable for water (not a single molecule has passed across the layer in a whole 500 ns of MD). The bl-2 and bl-3, which are considered to be in a liquid-crystalline state, let pass approximately two and one water molecule per 10 ns, respectively, which is 10- or 20-fold lower than phospholipid bilayers do. The bl-4 and bl-5 systems, which have only two branching modifications in hydrophobic tails, have very small, yet non-zero permeability. It‘s important to note that ring-containing lipids provide $\approx$ 2-fold lower water permeability as compared to corresponding methyl-containing mimetics (bl-3 vs. bl-2 and bl-4 vs. bl-5). This observation provides support for the fact that cyclopentane rings form an extra barrier for water and solutes to diffuse through the membrane.

3. Conclusion

Our MD results provide evidence that the unique high durability and low permeability of archaeal membranes, along with their liquid-crystalline state within a very broad temperature range, are determined by the chemical nature of isoprenoid tails themselves along with the branching groups, which are mostly side

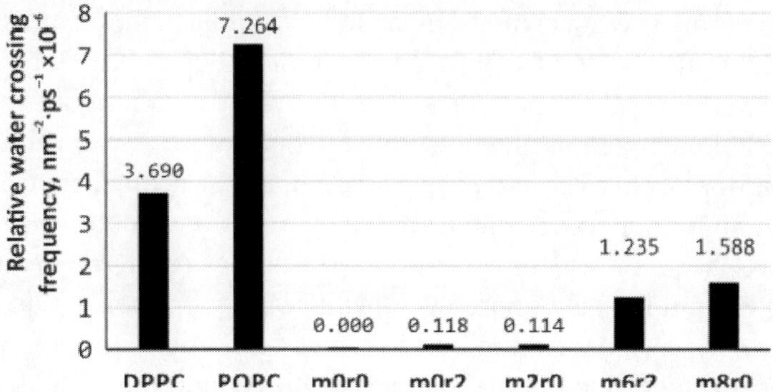

Figure 4. Permeation of water across the membranes. The measured quantity is frequency of water passage through the membrane (without normalization on MD box cross-section; DPPC/POPC membranes had approximately the same surface area). Simulation temperature is 350 K.

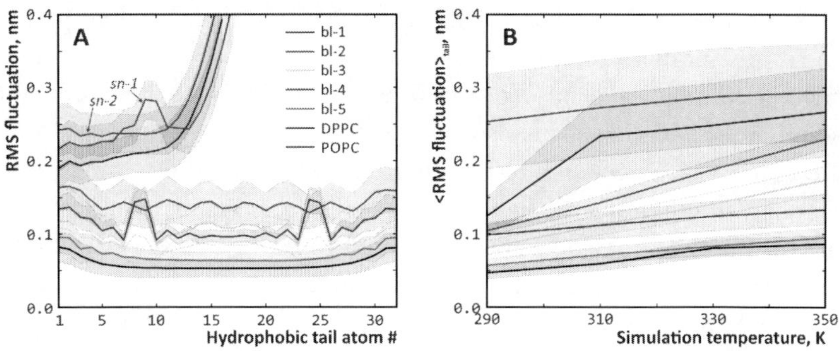

Figure 5. Bolalipid mimetics with branched tails are more liquid than straight-tailed ones. The measure of liquidity is MD-averaged RMS fluctuation (RMSF) of either certain tail atom (A) or the whole tail(B) (see Methods). A. RMSF profiles for five bolalipid mimetics, and also DPPC and POPC for comparison (see legend). Shaded areas are standard deviation. B. Tail-averaged RMSF (<RMSF>) at four simulation temperatures.

methyls and chain-incorporated cyclopentane rings. Apart from phospholipid bilayers, the low amount of fast-fluctuating void spaces inside the membrane ensures low permeability of the bolalipid layers for the solvent and the rather unique microscopic organization of their interface even with the PC polar heads inherent in convenient lipid membranes.

A thorough comparison of the atomistic-scale properties of archaeal-like membranes and common lipid bilayers reveals a very intriguing picture. Namely, due to the free ends of acyl chains, the hydrophobic core of lipid bilayers is rather dynamicand characterized by fast large-scale fluctuations, clustering and appearance of free volumes. The bilayer interface is dynamic as well (although to a lesser extent) and manifests a mosaic nature, mainly determined by the type of polar heads [23]. Such complex behavior is biologically relevant because it ensures a fast and adequate reaction of bacterial and eukaryotic membranes to a wide variety of scenarios of functioning, including adaptation to binding, insertion of peptides and proteins, membrane fusion and rearrangements, and so on. Despite these local perturbations, lipid bilayer membranes preserve their interfacial organization from external intrusions well. Water plays a crucial role in this. As we demonstrate here for the first time, the analogous characteristics of archaeal-like membranes exhibit a prominent dualism while some of them very closely resemble lipid bilayers, others are drastically different. For instance, the tight packing of lipid chains prevents the appearance of voids and fluctuations in the hydrophobic domain, although the mosaic nature and water dynamics in the interface can be efficiently regulated within a wide range, namely by the incorporation of different branching groups into the lipid tails. This allows for the creation of membranes with unique mechanical properties (durability, stability at high temperatures / pressure, etc.), but still biologically active. It is noteworthy that the response of bolalipids to temperature changes can also be very unusual, unlike in lipid membranes. These peculiarities of atomistic structure of the hydrophobic core of the membrane may have provided Archaea with the ability to thrive in the harshest environments. The extraordinary ability of bolalipid membranes to vary their physical properties depending on the number and chemical nature of side branching groups sheds some light on the next steps in the rational design of novel materials for biotechnology and medicine, based on archaeal-type lipids. Future theoretical works should pay attention on modeling of mechanistic properties of archaeal-like membranes: their response to pressure, deformation and formation of the pores, as well as on more systematic assessment of membranes permeability not only for water, but also for other solutes (e.g., ions) and gases. One should also note that due to the extraordinary properties of archaeal membranes, bolalipids will certainly provide some additional degrees of freedom in the construction of such materials. In other words, bolalipids significantly broaden the existing library of lipids used in development new membrane systems. These may be artificial membranes and vesicles with increased durability and carrying capacity, as well as unique surface properties that would define interaction with other agents, such as peptides and proteins.

4. Computational Methods

To study the effect of hydrophobic tails branching groups, five mimetics of archaeal tail-linked GDGT lipids were chosen. The length of the main tail chain was 32 carbon atoms, linked by ether bonds to archaeal-type glycerol backbone. To bring in focus the role of hydrophobic core chemical structure, polar heads were chosen to be just phosphatidylcholine. Mimetics differed by number and type of modifications of the main chain of the hydrophobic tail from complete absence of branching groups (system is referred to as bl-1) to introduction of eight modification per tail in bl-2 (eight methyls) and bl-3 systems (six methyls and two cyclopentane rings). In addition, there were two intermediate variants with either two rings (bl-4) or two methyls (bl-5). The position of methyl groups and rings were taken from natural lipids structure. The summary of these modifications in studied mimetics is given in Figure 1.

Molecular dynamics was performed with Gromacs 4.5 [27] using Slipids [28] force field. Bolalipids topologies were created by analogy with common lipids, considering all tail atoms non-charged. For each bolalipid, four temperatures (290, 310, 330 and 350 K) were studied; Nose-Hoover thermostat was used. Pressure was controlled in a semiisotropic manner with Parrinello-Rahman barostat. MD integration time was 0.002 ps; The van der Waals and electrostatic interactions were truncated using the twin range 1.0/1.4 nm spherical cut-off. Electrostatic effects were treated using PME scheme. The length of each trajectory was 500 ns.

Studied systems contained 144 lipid molecules each with approximately 30000 water molecules added. Starting configuration was created by translation of a single lipid to a 12x12 grid with a step of 1 nm (yielding 12x12x10 nm^3 MD box). Next, the water was added and subsequently removed from hydrophobic core area. Equilibration was performed in three steps: 1) 50 ps MD with fixed volume (to get rid of system construction defects); 2) 250 ps MD with semi-isotropic volume coupling scheme using Berendsen barostat and 3) 500 ps MD with semiisotropic volume coupling scheme using Parrinello-Rahman barostat for better lateral adjustment of lipid packing. To decrease computational error a 0.001 ps integration time step was used during equilibration. After that, production MD runs were performed.

To assess lipid packing and phase condition of the model membranes, each bolalipid was represented as a vector built with two sn-2 glycerol atoms from each polar head (Fig.2A). Average lipid tilt angle χ, average lipid-lipid collinearity angle φ and average lipid length L (Fig.2A; Fig. 3A-C) were calculated with in-house scripts to assess packing characteristics. Partial density profiles (Fig. 3D) was calculated with g_density program from the Gromacs suite. From these profiles, membrane thickness (as a distance between two phosphorus peaks) and hydrophobic core thickness (as 99%-drop of solvent density) were calculated.

Acknowledgments

A.O.C., I.A.B. and P.E.V. express their gratitude to the Russian Foundation for Basic Research (grants 13-04-40326- and 13-04-00825) for support. This work was supported by Russian Scientific Foundation (grant 14-14-00871 awarded to R.G.E.). A.O.C. is recipient of the Stipend of the President of the Russian Federation. Authors also thank RAS Programmes Basic fundamental research for nanotechnologies and nanomaterials and Molecular and Cellular Biology. Access to computational facilities of the Joint supercomputer center of RAS (Moscow) is greatly appreciated.

References

[1] Offre, P.; Spang, A.; Schleper, C., Archaea in biogeochemical cycles. *Annual Review of Microbiology*, 2013, 67, 437-57.

[2] Goldenfeld, N., Looking in the right direction: Carl Woese and evolutionary biology. *RNA Biology*, 2014, 11.

[3] Woese, C. R.; Kandler, O.; Wheelis, M. L., Towards a natural system of organisms: proposal for the domains Archaea, Bacteria, and Eucarya. *Proceedings of the National Academy of Sciences of the United States of America*, 1990, 87, 4576-9.

[4] Koga, Y., From Promiscuity to the Lipid Divide: On the Evolution of Distinct Membranes in Archaea and Bacteria. *Journal of Molecular Evolution*, 2014.

[5] Oger, P. M.; Cario, A., Adaptation of the membrane in Archaea. *Biophysical Chemistry*, 2013, 183, 42-56.

[6] Lombard, J.; Lopez-Garcia, P.; Moreira, D., Phylogenomic investigation of phospholipid synthesis in archaea. *Archaea 2012*, 2012, 630910.

[7] Chong, P. L.; Ayesa, U.; Daswani, V. P.; Hur, E. C., On physical properties of tetraether lipid membranes: effects of cyclopentane rings. *Archaea 2012*, 2012, 138439.

[8] Chong, P. L.; Sulc, M.; Winter, R., Compressibilities and volume fluctuations of archaeal tetraether liposomes. *Biophysical Journal*, 2010, 99, 3319-26.

[9] Bartucci, R.; Gambacorta, A.; Gliozzi, A.; Marsh, D.; Sportelli, L., Bipolar tetraether lipids: chain flexibility and membrane polarity gradients from spin-label electron spin resonance. *Biochemistry*, 2005, 44, 15017-23.

[10] Chong, P. L.; Ravindra, R.; Khurana, M.; English, V.; Winter, R., Pressure perturbation and differential scanning calorimetric studies of bipolar tetraether liposomes derived from the thermoacidophilic archaeon Sulfolobus acidocaldarius. *Biophysical Journal*, 2005, 89, 1841-9.

[11] Elferink, M. G.; de Wit, J. G.; Driessen, A. J.; Konings, W. N., Stability and proton-permeability of liposomes composed of archaeal tetraether lipids. *Biochimica et Biophysica Acta*, 1994, 1193, 247-54.

[12] Dante, S.; Rosa, M. D.; Maccioni, E.; Morana, A.; Nicolini, C.; Rustichelli, F.; Troitsky, V. I.; Yang, B., Thermal Stability of Bipolar Lipid Langmuir Blodgett Films by X-Ray Diffraction. *Molecular Crystals and Liquid Crystals Science and Technology. Section A. Molecular Crystals and Liquid Crystals*, 1995, 262, 191-207.

[13] Chong, P. L., Archaebacterial bipolar tetraether lipids: Physico-chemical and membrane properties. *Chemistry and physics of lipids*, 2010, 163, 253-65.

[14] Boyd, E. S.; Hamilton, T. L.; Wang, J.; He, L.; Zhang, C. L., The role of tetraether lipid composition in the adaptation of thermophilic archaea to acidity. *Frontiers in Microbiology*, 2013, 4, 62.

[15] Zhai, Y.; Chong, P. L.; Taylor, L. J.; Erlkamp, M.; Grobelny, S.; Czeslik, C.; Watkins, E.; Winter, R., Physical properties of archaeal tetraether lipid membranes as revealed by differential scanning and pressure perturbation calorimetry, molecular acoustics, and neutron reflectometry: effects of pressure and cell growth temperature. Langmuir : *The ACS Journal of Surfaces and Colloids*, 2012, 28, 5211-7.

[16] Nishihara, M.; Koga, Y., Extraction and composition of polar lipids from the archaebacterium, Methanobacterium thermoautotrophicum: effective extraction of tetraether lipids by an acidified solvent. *Journal of Biochemistry*, 1987, 101, 997-1005.

[17] Dannenmuller, O.; Arakawa, K.; Eguchi, T.; Kakinuma, K.; Blanc, S.; Albrecht, A. M.; Schmutz, M.; Nakatani, Y.; Ourisson, G., Membrane properties of archaeal macrocyclic diether phospholipids. *Chemistry*, 2000, 6, 645-54.

[18] Stewart, L. C.; Kates, M.; Ekiel, I. H.; Smith, I. C. P., Molecular order and dynamics of diphytanylglycerol phospholipids: a 2H and 31P-NMR study. *Chemistry and Physics of Lipids*, 1990, 54, 115-129.

[19] Benvegnu, T.; Lemigre, L.; Cammas-Marion, S., Archaeal Lipids: Innovative Materials for Biotechnological Applications. *European Journal of Organic Chemistry*, 2008, 2008, 4725-4744.

[20] Bulacu, M.; Periole, X.; Marrink, S. J., In silico design of robust bolalipid membranes. *Biomacromolecules*, 2012, 13, 196-205.

[21] Gabriel, J. L.; Chong, P. L., Molecular modeling of archaebacterial bipolar tetraether lipid membranes. *Chemistry and Physics of Lipids*, 2000, 105, 193-200.

[22] Shinoda, W.; Shinoda, K.; Baba, T.; Mikami, M., Molecular dynamics study of bipolar tetraether lipid membranes. *Biophysical Journal*, 2005, 89, 3195-202.

[23] Pyrkova, D. V.; Tarasova, N. K.; Krylov, N. A.; Nolde, D. E.; Pentkovsky, V. M.; Efremov, R. G., Dynamic clustering of lipids in hydrated two-component membranes: results of computer modeling and putative biological impact. *Journal of Biomolecular Structure & Dynamics*, 2013, 31, 87-95.

[24] Pyrkova, D. V.; Tarasova, N. K.; Pyrkov, T. V.; Krylov, N. A.; Efremov, R. G., Atomic-scale lateral heterogeneity and dynamics of two-component lipid bilayers composed of saturated and unsaturated phosphatidylcholines. *Soft Matter*, 2011, 7, 2569.

[25] Franck, J. M.; Scott, J. A.; Han, S., Nonlinear scaling of surface water diffusion with bulk water viscosity of crowded solutions. *Journal of the American Chemical Society*, 2013, 135, 4175-8.

[26] Krylov, N. A.; Pentkovsky, V. M.; Efremov, R. G., Nontrivial behavior of water in the vicinity and inside lipid bilayers as probed by molecular dynamics simulations. ACS nano, 2013, 7, 9428-42.

[27] Hess, B.; Kutzner, C.; van der Spoel, D.; Lindahl, E., GROMACS 4:Algorithms for Highly Efficient, Load-Balanced, and Scalable Molecular Simulation. *Journal of Chemical Theory and Computation*, 2008, 4, 435-447.

[28] Jambeck, J. P.; Lyubartsev, A. P., Derivation and systematic validation of a refined all-atom force field for phosphatidylcholine lipids. *The Journal of Physical Chemistry B*, 2012, 116, 3164-79.

In: Computational Materials ... ISBN: 978-1-63482-541-2
Editor: K. T. Kholmurodov, pp. 97-108

Chapter 10

The Structural Behavior of the DNA and CPD in a Repairing Molecular System

Ermuhammad B. Dushanov[1,2,*] ***and Kholmirzo T. Kholmurodov***[3,4]
[1]Laboratory of Radiation Biology,
Joint Institute for Nuclear Research,
Dubna, Moscow Region, Russia
[2]Institute of Nuclear Physics,
Ulugbek, Tashkent, Uzbekistan
[3]Laboratory of Neutron Physics,
Joint Institute for Nuclear Research,
Dubna, Moscow Region, Russia
[4]Dubna International University,
Dubna, Moscow Region, Russia

Abstract

The subject of this work is to study the behavior of the DNA molecule in a repairing molecular system. For a repairing molecular system we investigate the *Anacystis nidulans* photolyase enzyme that contains two chromophore cofactors: flavin adenine dinucleotide (FAD; flavin) and 8-hydroxy-5-deazaflavin (8-HDF, deazaflavin). This photolyase enzyme was simulated for a photoreaction DNA repairing process via molecular

*E-mail address: dushanov@jinr.ru

dynamics (MD) method. Observations show that flavin has actively participated in the interaction process thereby involving the conformational changes of the DNA and CPD substrate fragments. We have observed a distinct behavior between two types of DNA Conformations, consequently matching the average characteristics of a healthy DNA with its thymine. This gives us a basis to assume that the damaged DNA, under the influence of the photolyase enzyme' active cofactor, has to rotate around its symmetry axis and thereby tending to flip out its CPD fragment.

Keywords: Photolyase Enzyme, Flavin, Deazaflavin, DNA Damage, Cofactors

1. Introduction

Photolyases (PHTs) are unique DNA repair enzymes (monomeric protein of 450-550 amino acids) that as light-dependently catalyzing agents serve in a conversion mechanism of the damaged photoproducts (PPs) to an original state. For today the three crystal structures of the DNA PHTs are known [1, 2, 3]. There are two types of the DNA damages: cyclobutane pyrimidine dimers (CPDs) and (6-4) photoproducts ((6-4) PPs) [4]. (We imply thymine dimers, Pyr<>Pyr, for the CPDs, and Pyr[6-4]Pyr for pyrimidine-pyrimidone (6-4) PPs). The PHTs repair the CPD and (6-4) PPs, herewith recovering back the damaged DNA to its original (native) state. The mechanism of the CPD lesion recognition and binding have been described in the literature throughout the biochemical data [5, 6], computer modeling [1, 7, 8], nuclear magnetic resonance (NMR) spectroscopy [9] and so on. According to these data the PHTs flip the damaged di-nucleotide out of the DNA double helix into its own active site, and next perform a substrate binding. Following the photon absorbing through its antenna the PHT pigment triggers the excitation energy transfer to a catalytic cofactor which is called a FAD (flavin adenine di-nucleotide) (see Fig. 1). Between the catalytic active FAD (shortly, flavin) and CPD the exchange of an excess energy for the repairing of the damaged di-nucleotide occurs. The CPD repairment thus results into splitting of two pyrimidines to their original shapes. As a final reaction step, the electron charge will be transferred back into flavin, thereby restoring a flavin's catalytically active form.

Described above the DNA repairing mechanism works in certain organisms that possess a photoreactivating PHT enzyme inside. The enzyme contains two chromophores which are capable to capture photons from the UV-B (ultraviolet

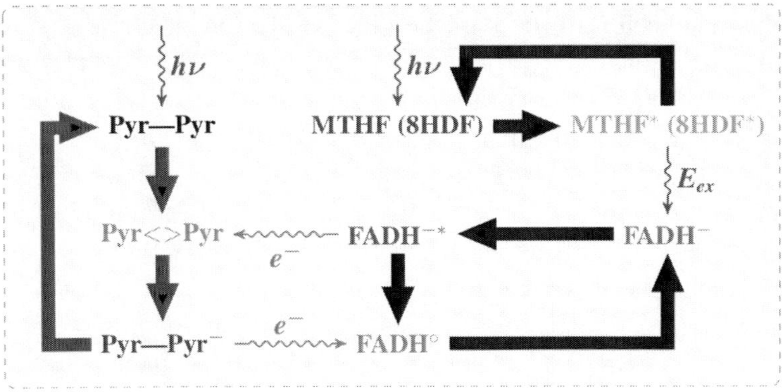

Figure 1. A reaction mechanism of a blue light-mediated repair of CPD lesion by the DNA PHTs (photolyases). The UV influence damages the DNA neighboring pyrimidines (Pyr–Pyr) and induces a CPD (Pyr<>Pyr) lesion. The cofactor (MHTF or 8-HDF) adsorbs light photon and transfers it to a catalytically active flavin ($FADH^-$) as an excitation energy E_{ex} via dipole–dipole interaction. The flavin molecule ($FADH^{-*}$) transfers an electron to CPD. The CPD splits out into two pyrimidines, thereby transferring back the electron to a newly formed FADH°; in its final state the flavin molecule has to pass into a catalytically active form.

blue) light of the wavelength between 350-450 nm. The PHT detects, recognizes and binds the damaged DNA (a DNA's CPD site). The light energy in a visible range (350-450 nm) is being absorbed by a light-harvesting cofactor 5,10-Methylenetetrahydrogolate (MTHF) or 8-hydroxy-5-deazaflavin (8-HDF). The adsorbed energy is used in the oxidation of a cyclobutane ring, and, subsequently, in conversion of the thymine's dimmer into two separate monomers. As a result, the PHT enzyme dissociates from the DNA, therefore the damage has repaired.

It is worth noting that the CPD recognition process by the PHT does not influence the content of the DNA nucleotides. More details of the DNA repair by PHTs could be found in [10] where the PHT mechanism has been found in prokaryotes, lower and higher eukaryotes. Nevertheless, the PHT existence in placental mammals is still unknown.

The CPD repair mechanism in human takes place through the excision repair which occurs in the dark conditions [10, 11]. The excision repair does not require light; the PHT; instead of the repairing of breaking bonds it only recognizes where the region of damaged nucleotides is located. The protein complex recognizes the distortion of the DNA molecule (CPDs), so that the related DNA strand with its CPD to be cutted off. Generally, 12 nucleotides could be recognized and a DNA part with its marked chain could be removed. A DNA polymerase-I [12] fills up the gap and a DNA ligase [13] removes the recognized part. Other DNA repair mechanism includes mutagenic repair or dimer bypass, a recombination repair, cell-cycle checkpoints, apoptosis and certain alternative repair pathways and so on. The above mentioned mechanism can operate in various organisms that involve the CPDs or (6-4) PPs correction process [14]. The DNA repair process has also been reviewed in [15], where the crystal structures of the replicative and bypass DNA polymerases were encountered. Thereby the two typical lesions arising from the oxidation of DNA were distinguished.

The DNA UV damages are generally responsible for the induction of skin tumors, the formation of cell carcinomas and melanomas [16]. The UV induced dimmers for the DNA can brought into a mispairing of the DNA strands, so far, the induced damage in the next stage of the DNA copying can stop a precise programmed replication altogether. Unless the DNA damage is repaired the pyrimidine dimers can lead to a transcription blocking, to the various mutations or, finally, to a cell death, thereby originating a cancer disease [17].

2. Structure and Model

In our previous papers [18, 19] we have discussed the charge transfer mechanism from the active cofactor to a CPD fragment before the repairing and getting back on the process finishing. In the present study we simulate the deoxyribodipyrimidine PHT protein *A. nidulans* and examine the DNA behavior with a flipped-in CPD. All properties were compared with those of the normal (repaired) DNA ones. The structural model for the photolyase system with the two cofactors (FAD and 8-HDF) was built using the *A. nidulans* complex (PDB entry 1TEZ) as a template.

Figs. 2 demonstrates the *A. nidulans* crystal structure where a ribbon diagram representation includes the α-helices, β-strands and 3_{10}-terminals in different colors. We have introduced several structures of the flipped-out CPD-like DNA, so to be able to investigate the damaged and repaired states of the DNA molecule by the PHT enzyme alltogether.

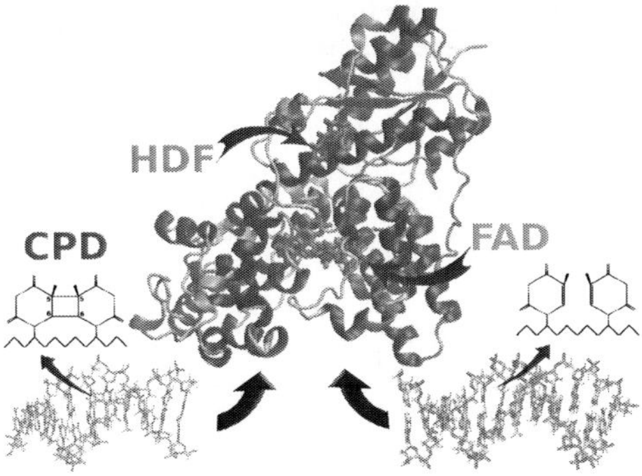

Figure 2. A crystal structure of the *A. nidulans* photolyase and different DNAs. The β-strands are colored in magenta, the α-helices are in blue and the 3_{10}-helices are in green. Depicted on the left the DNA molecule is damaged and it contains flipped-in CPD substrate. The flavin and deazaflavin cofactors are shown in red color within the photolyase protein.

Pyr<>Pyr	Pyr–Pyr
5′ -- GCTTAATTCG -- 3′	5′ -- CGCGAATTTCGCG -- 3′
3′ -- CGAATTAAGC -- 5′	3′ -- GCGCTTAAGCGC -- 5′

Figure 3. Flipped-in CPD and repaired DNA nucleotides sequences.

The nucleotide sequences of the simulated DNA structures are shown in Fig. 3.

For performing the MD simulation we have used the Amber 11 software package with AmberTools 1.5 module [20]. We built up two models including various DNA structures: DNA with a flipped-in CPD and normal DNA with the two splitted thymine pairs. The flipped-in CPD has a neutral charge (-2). At the same time, the conditions for (-1) charged FAD remain the same between two considered models. We imply the $FADH^{-*}$ for the FAD molecule with an activated excitation energy and the $FADH^{-}$ for a normal state of the FAD molecule.

The MD calculations for two molecular configurations were completed on 3 ns dynamical changes. The flavin and CPD structures were restrained; the weights of the positional restraints varied from 0.0 to 0.005 kcal/mol·Å. The Langevin dynamics temperature regulation with the parameters T0 = 300 K, IG

= 71277 and the collision frequency $\gamma = 2$ ps^{-1} [21, 22, 23] were used. For the bond length constraints the SHAKE algorithm [24] with the parameter NTC = 2 was used for the default relative geometrical tolerance of coordinate resetting.

3. Results and Discussions

For the PHT-DNA interaction process the two molecular configurations were simulated and their comparative analysis was performed. Note that for such interactions there are no exchange of the charges. In the first model the two adjusted thymines were connected together (damaged DNA), in the second model the adjusted thymines were represented as a separate states (repaired DNA). During the DNA repairing process we do have both flavin and CPD substrate as charged substances. In this process the flavin fragment is activated by a some excitation energy that causes transferring of the electron to the CPD. This charge transfer plays a key role in the DNA reparing. After getting the two thymines back into a splitted states there a return process has to manifest — the CPD transfers its electron back into the flavin [18].

3.1. Damaged DNA Model

As a first approach our interest was concentrated on the flavin ($FADH^{-*}$) and flipped-in CPD (Pyr<>Pyr) interaction dynamics. Simulations with the first configuration as described above observe that the interaction of these two components has to result on the CPD conformational changes from flipped-in to flipped-out. The MD results for such conformations are shown in Fig. 4. The left diagram of Fig. 4 shows the RMSD (root mean-square displacements) for the CPD, flavin (FAD) and DNA, respectively. In the right diagram of Fig. 4 the distance calculation results illustrating the CPD conformational dynamics are presented. The distance diagrams were constructed between the one of the CPD thymine and an opposite adenine nucleotide.

Flavin as an essentially mobile molecule that periodically oscillates around its equilibrium position. Flavin's motion eventually influences the CPD conformational behavior. For example, the flavin-CPD interaction process results in the DNA rotation around its symmetrcal axis. The RMSD for the damaged DNA shows a gradually increasing from 2 to 5 Å. Nevertheless, the CPD conformation in this model do not change esssentially. The distance between two nucleotides increases at t = 1.0 ns from 4 to 15 Å. This is due to that the

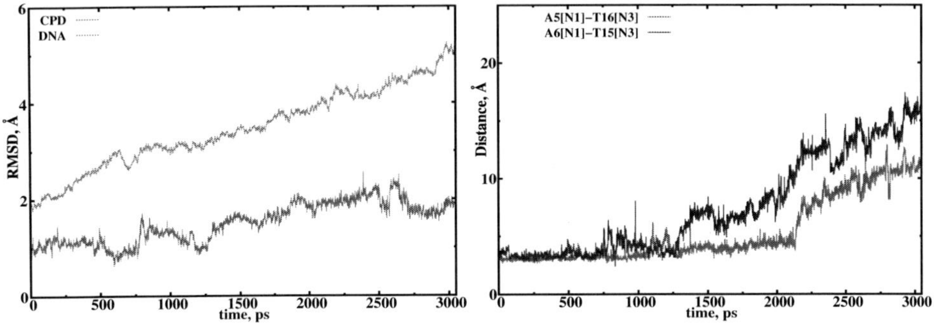

Figure 4. The RMSD and distance for the PHT-CPD model. On the left, the RMSD are presented for the flipped-in CPD (red) and DNA (magenta). On the right, the interatomic distances are presented for the nucleotide pair thymine-adenine.

CPD substrate, under the influence of the flavin cofactor, will take a flipped-out conformation shape.

3.2. Full-Repaired DNA Model Dynamics

The mobile FAD molecule do not significantly changes the conformation of the DNA and repaired thymines.

In Fig. 5 the MD results are presented for a second model where the configuration of two repaired thymines (Pyr–Pyr) has seen to posses not an active interaction with the flavin molecule. As it was noted above, there is no electron transfer among the DNA, CPD substrate and FAD molecule. So far, in the MD simulations we have observed that both thymines are rotate inward the DNA structure to be hidden in it. This process indicate that the repaired DNA behaves as being as a normal one.

The repaired DNA behavior is different from that of the damaged DNA with a flipped-in CPD. The RMSD of the DNA molecule has slowly increases from the 1.5 up to 3 Å. The RMSD for the repaired thymine is close to the one of a normal DNA. The distance between the nucleotide pair changes from 4 to 10 Åso that the flavin influence has to be minimal.

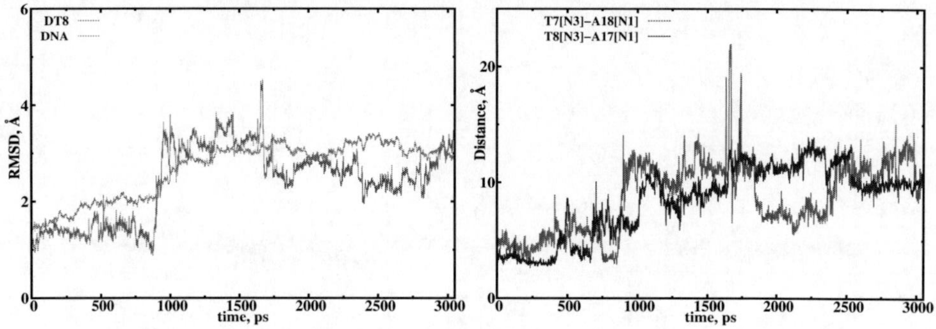

Figure 5. The RMSD and distance for the PHT-repaired DNA model. On the left, the RMSD are presented for the thymine under residue eight (red) and DNA (magenta). On the right, the interatomic distances are presented for the nucleotide pair thymine-adenine.

4. Conclusion

The present study as we considered above may serve as an useful approach for the understanding of the DNA repair mechanism and the nature of conformational changes in the subsiquent UV photoproducts. The MD simulations with *A. nidulans* enzyme provide a new glance on the DNA repair mechanism which can easily be extended for the FAD and CPD interactions, specifically, before and after repair processes.

Dynamics in the CPD repair is the interaction between the flavin ($FADH^{-*}$) and CPD (Pyr<>Pyr, where two adjusted thymines are binded together), described by the model with a flipped-in CPD. Simulations with such configuration observe that the interaction of these two components, flavin and CPD, and the behavior of flavin–CPD exchange are the predominant ones. Flavin being as an essentially mobile molecule oscillates fast around its equilibrium position, thereby influencing the CPD conformational behavior. For the initial state the flavin-CPD exchange results on a DNA rotation around a symmetry axis; the DNA conformation in this model does not changes visibly. However, the CPD conformation undergoes a strong influence as results of the interaction with flavin.

Dynamics after CPD repair is the interaction between the flavin ($FADH^{-}$) and thymine nucleotides (Pyr–Pyr, where two adjusted thymines are separate from each other), described by a second model. The DNA configuration in this model exhibits a clear repulsive nature of the two adjusted thymine. The interaction between these two adjusted thymines observe a rotation inward the DNA structure. Such process looks like as how repaired DNA structure behaves into a normal state.

In conclusion, the DNA-PHT structural changes could be studied in some detail through the atomistic/molecular simulation as one employed in the present study. Until now for such complex multiparticle system, as PHT enzyme interacting with DNA, performing more precise quantum chemistry calculations seem to be nontrivial and challenging task. The DNA conformations are essentially influenced by the charge distributions between the DNA and PHT. In the present study, the proposed MD models maintain a main feature of the DNA-PHT conformations resulting from peculiarities of the charge distribution scenarios in the system.

References

[1] H.W. Park, S.T. Kim, A. Sancar, and J. Deisenhofer. Crystal structure of DNA photolyase from *Escherichia coli*. *Science*, 268(5219), p.1866–1872, 1995.

[2] Tamada T, Kitadokoro K, Higuchi Y, Inaka K, Yasui A, de Ruiter PE, Eker AP, and Miki K. Crystal structure of DNA photolyase from *Anacystis nidulans*. *Nature Structural Biology*, 4(11), p. 887–891, 1997. Correspondence.

[3] Hirofumi Komori, Ryoji Masui, Seiki Kuramitsu, Shigeyuki Yokoyama, Takehiko Shibata, Yorinao Inoue, and Kunio Miki. Crystal structure of thermostable DNA photolyase: Pyrimidine-dimer recognition mechanism. *Proceedings of the National Academy of Sciences*, 98(24):13560–13565, 2001.

[4] Masato Kondoh, Kenichi Hitomi, Junpei Yamamoto, Takeshi Todo, Shigenori Iwai, Elizabeth D. Getzoff, and Masahide Terazima. Light-induced conformational change and product release in DNA repair by (6?4) photolyase. *Journal of the American Chemical Society*, 133(7), p.2183–2191, 2011.

[5] Kathleen S. Christine, Alexander W. MacFarlane, Kongsheng Yang, and Robert J. Stanley. Cyclobutylpyrimidine dimer base flipping by DNA photolyase. *Journal of Biological Chemistry*, 277(41), p.38339–38344, 2002.

[6] Brian J. Vande Berg and Gwendolyn B. Sancar. Evidence for dinucleotide flipping by DNA photolyase. *Journal of Biological Chemistry*, 273(32), p.20276–20284, 1998.

[7] Don B. Sanders and Wiest Olaf. A model for the enzyme?substrate complex of DNA photolyase and photodamaged DNA. *Journal of the American Chemical Society*, 121(22), p.5127–5134, 1999.

[8] Jens Antony, Dmitry M. Medvedev, and Alexei A. Stuchebrukhov. Theoretical study of electron transfer between the photolyase catalytic cofactor $FADH^-$ and DNA thymine dimer. *Journal of the American Chemical Society*, 122(6), p.1057–1065, 2000.

[9] Takuya Torizawa, Takumi Ueda, Seiki Kuramitsu, Kenichi Hitomi, Takeshi Todo, Shigenori Iwai, Kosuke Morikawa, and Ichio Shimada. Investigation of the cyclobutane pyrimidine dimer (CPD) photolyase DNA recognition mechanism by NMR analyses. *Journal of Biological Chemistry*, 279(31), p.32950–32956, 2004.

[10] Rajeshwar P. Sinha and Donat-P. Hader. UV-induced DNA damage and repair: a review. *Photochem. Photobiol. Sci.*, 1, p.225–236, 2002.

[11] Aziz Sancar. Structure and function of DNA photolyase and cryptochrome blue-light photoreceptors. *Chemical Reviews*, 103(6), p.2203–2238, 2003.

[12] Lehman I.R., Bessman M.J., Simms E.S., and Kornberg A. Enzymatic synthesis of deoxyribonucleic acid. I. Preparation of substrates and partial purification of an enzyme from *Escherichia coli*. *J. Biol. Chem.*, 233(1), p.160–170, 1958.

[13] I. R. Lehnman. DNA ligase: Structure, mechanism, and function. *Science*, 186(4166), p.790–797, 1974.

[14] Vanessa Chigancas, Eliane Namie Miyaji, Alysson Renato Muotri, Jacqueline de Fátima Jacysyn, Gustavo Pessini Amarante-Mendes, Akira Yasui, and Carlos Frederico Martins Menck. Photorepair prevents

ultraviolet-induced apoptosis in human cells expressing the marsupial photolyase gene. *Cancer Research*, 60(9), p.2458–2463, 2000.

[15] Karl E Zahn, Susan S Wallace, and Sylvie Doublié. DNA polymerases provide a canon of strategies for translesion synthesis past oxidatively generated lesions. *Current Opinion in Structural Biology*, 21(3), p.358 – 369, 2011.

[16] Nerys R. Morse, Valerie Meniel, and Raymond Waters. Photoreactivation of UV-induced cyclobutane pyrimidine dimers in the MFA2 gene of Saccharomyces cerevisiae. *Nucleic Acids Research*, 30(8), p.1799–1807, 2002.

[17] Bernhard Suter, Magdalena Livingstone-Zatchej, and Fritz Thoma. Chromatin structure modulates DNA repair by photolyase *in vivo*. *EMBO J*, 16(8), p.2150–2160, 1997.

[18] Ermuhammad Dushanov and Kholmirzo Kholmurodov. A novel approach to simulate a charge transfer in DNA repair by an *Anacystis nidulans* photolyase. *The Open Biochemistry Journal*, 8, p.35 – 43, 2014.

[19] Kholmirzo Kholmurodov, Ermuhammad Dushanov and Kenji Yasuoka. Molecular dynamics simulations of a DNA photolyase protein: High-mobility and conformational changes of the FAD molecule a low temperatures. *Advances in Bioscience and Biotechnology*, 3, p.169 – 180, 2012.

[20] David A. Case, Thomas E. Cheatham, Tom Darden, Holger Gohlke, Ray Luo, Kenneth M. Merz, Alexey Onufriev, Carlos Simmerling, Bing Wang, and Robert J. Woods. The Amber biomolecular simulation programs. *Journal of Computational Chemistry*, 26(16), p.1668–1688, 2005.

[21] Richard W. Pastor, Bernard R. Brooks, and Attila Szabo. An analysis of the accuracy of Langevin and molecular dynamics algorithms. *Molecular Physics*, 65(6), p.1409–1419, 1988.

[22] Richard J. Loncharich, Bernard R. Brooks, and Richard W. Pastor. Langevin dynamics of peptides: The frictional dependence of isomerization rates of N-acetylalanyl-N'-methylamide. *Biopolymers*, 32(5), p.523–535, 1992.

[23] Jesús A. Izaguirre, Daniel P. Catarello, Justin M. Wozniak, and Robert D. Skeel. Langevin stabilization of molecular dynamics. *The Journal of Chemical Physics*, 114(5), p.2090–2098, 2001.

[24] Jean-Paul Ryckaert, Giovanni Ciccotti, and Herman J.C Berendsen. Numerical integration of the cartesian equations of motion of a system with constraints: molecular dynamics of n-alkanes. *Journal of Computational Physics*, 23(3), p.327 – 341, 1977.

In: Computational Materials ... ISBN: 978-1-63482-541-2
Editor: K. T. Kholmurodov, pp. 109-121

Chapter 11

LOCAL ENVIRONMENT ANALYSIS OF MAGNETIC ELEMENTS IN FUNCTIONAL MATERIALS BY X-RAY ABSORPTION NEAR-EDGE STRUCTURE MEASUREMENTS AND FIRST-PRINCIPLES CALCULATIONS

Tomoyuki Yamamoto[1,2,*], ***Kazuma Nishimura***[1], ***and Tsuyoshi Yoshioka***[1]
[1]Faculty of Science and Engineering
Waseda University,
Shinjuku, Tokyo, Japan
[2]Institute of Condensed-Matter Science,
Waseda University
Shinjuku, Tokyo, Japan

Abstract

The analytical method of local environment of dilutely doped magnetic elements in functional materials in an atomic scale by the X-ray absorption near-edge structure (XANES) measurements and the first-principles calculations is demonstrated. Results for the lead-free piezoelectric material, Mn and Fe doped $Bi_4Ti_3O_{12}$, is shown as an

*E-mail address: tymmt@waseda.jp

application of the above method.

Keywords: Local Environment Analysis, X-Ray Absorption Near Edge Structure, First-Principles Calculation, Functional Materials, Magnetic Element

1. Introduction

Doping technique, i.e., incorporation of the additional elements into matrix, has been widely used to add the new properties, e.g., magnetic, electric and optical properties, etc., to the materials. Such strategy has brought us a great success in many fields of the materials science. These so-called "functional materials" developed with doping technique are now in use as many kinds of industrial materials. Most of the studies for functional materials have been devoted into optimizing newly appeared properties by changing concentrations of dopants and/or matrix materials. In order to understand the influence of such doping on the newly appeared properties, it is essential to know the local environment of doped elements in an atomic scale. However, such analysis have not been always carried out. This is because the effective amount of doping is often very dilute, which leads us to a difficulty to determine the local environment of dopants. For such purpose, combined use of X-ray absorption near-edge structure measurements and first-principles calculations is quite powerful, then we have applied this method for the analysis of dilute dopants in wide varieties of functional materials [1, 2, 3]. In the current report, after the explanation of an analytical method of local environment of magnetic elements in functional materials, local environment analysis of magnetic elements, Mn and Fe, in lead-free piezoelectric material, $Bi_4Ti_3O_{12}$, is demonstrated [4] as one of the applications of our analytical method.

2. Analytical Method of Local Environment of Dilute Magnetic Elements in Functional Materials

In this section, our strategy for the local environment analysis of dilute magnetic 3d transition elements in the functional materials by using XANES and first-principles calculations is explained,which is schematically shown in Fig. 1. At first, we synthesize the materials of interest by doping technique changing the concentration of dopants and/or matrix materials. Then, the X-ray diffraction

(XRD) patterns are measured to determine the crystal structure of the materials and to check whether secondary phase exists or not. In this XRD measurements, we can also examine a change in lattice constants which helps to determine the substituted site of dopants. When the concentration of the dopants is large enough, the lattice parameter would change due to a difference in ionic radii between the ions in matrix to be substituted and doped ions. However, no change is often observed in XRD patterns due to the doping, since a concentration of dopant is often tiny. After such basic crystal structure analysis by XRD, samples are provided to the X-ray absorption near-edge structure (XANES) measurements. For the valence state analysis of 3d transition elements, L-edge XANES measurement is quite effective. This is because we can determine the valence state of 3d transition elements from the profile of the observed L-edge XANES. This analysis is a kind of so-called finger-print type analysis, then we just need the comparison with standard spectra. For the local geometrical structure analysis around the dopant, K-edge XANES is also observed, which is more sensitive to change in local environment, because K-edge XANES profile of the 3d transition elements is a reflection of unoccupied 4p partial density of states of the element of interest. However, for the analysis of K-edge XANES, finger-print type analysis is not appropriate, because K-edge XANES profiles are quite sensitive to change in local environment. Even if the small change appeares in the local environment of dopant of interest, XANES spectrum often changes drastically. Then the first-principles calculations using density functional theory (DFT) is employed to prepare the theoretical finger-prints for the analysis of K-edge XANES. Combining above 4 methods, 1) XRD, 2)L-edge XANES, 3) K-edge XANES and 4) first-principles DFT calculations, we can determine the local environment of dilute dopant.

3. Experimental Procedures

Mn and Fe doped $Bi_4Ti_3O_{12}$ were prepared by the solid state reaction method changing the concentration of Mn and Fe ions, i.e., x = 0.00, 0.01, 0.03 and 0.05 in $Bi_4(Ti_{1-x}TM_x)_3O_{12}$ where TM = Mn or Fe. Our raw materials were reagent-grade Bi_2O_3, TiO_2, α-Fe_2O_3 and Mn_2O_3. We mixed these powders in an agate mortar, then calcined them at 1023 K for 5 h in air. These calcined powders were ground and mixed again, and pressed into a cylindrical pellet form (10 mmϕ). We sintered these powder aggregates at 1223 K for 5 h in air. The crystal structures of all samples were verified using the powder X-ray

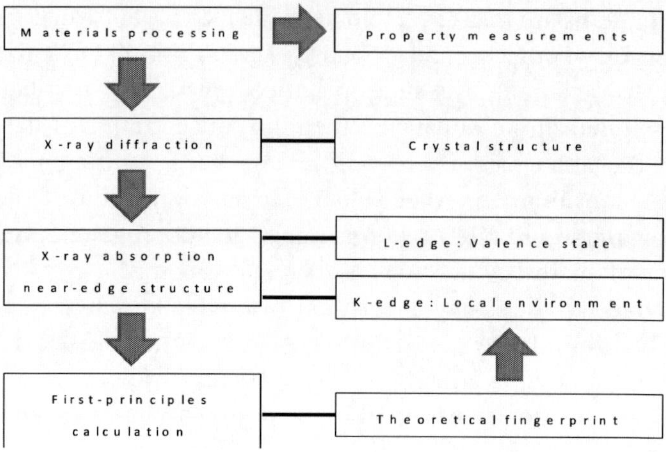

Figure 1. Flow of local environment analysis of dilute magnetic 3d transition element by XANES and first-principles calculations.

diffraction (XRD) technique with Cu-Kα X-rays.

Mn- and Fe-$L_{2,3}$ XANES spectra were measured at BL-4B in UVSOR, Okazaki, Japan, using the total electron yield method. The X-rays from the storage ring were monochromatized by a varied-line-spacing plane grating (800 lines/mm), and we set the energy resolution of the incident beams (E/ΔE) to 3000 by tuning the slit height at the upper and lower reaches of the grating. Mn- and Fe-K XANES spectra were collected at BL01B1 in SPring-8, Harima, Japan, using a transmission method. The incident beam from the storage ring was monochromatized by a Si(111) double-crystal monochromater. All the XANES measurements were carried out at room temperature.

4. Results and Discussion

The XRD patterns collected from the $Bi_4(Ti_{1-x}Mn_x)_3O_{12}$ and $Bi_4(Ti_{1-x}Fe_x)_3O_{12}$ specimens are shown in Fig. 2(a) and (b), respectively, together with the established XRD pattern of orthorhombic $Bi_4Ti_3O_{12}$ (JCPDS 35-0795). All the diffraction peaks could be indexed as orthorhombic $Bi_4Ti_3O_{12}$, and no extra peaks were found in the collected XRD patterns. The lattice parameters of our samples were calculated by using the conventional

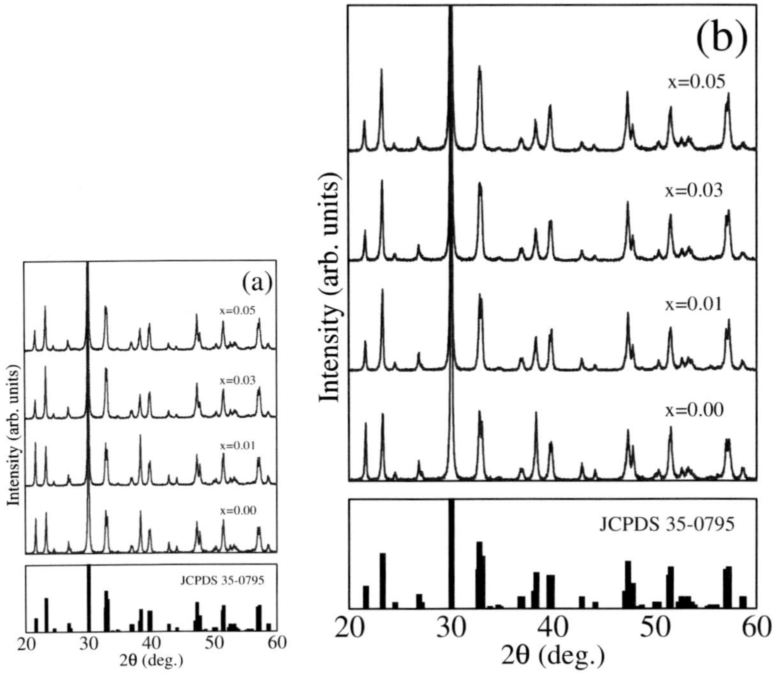

Figure 2. Observed XRD patterns of (a) $Bi_4(Ti_{1-x}Mn_x)_3O_{12}$ and (b) $Bi_4(Ti_{1-x}Fe_x)_3O_{12}$. Calculated XRD pattern of $Bi_4Ti_3O_{12}$ (JCPDS 35-0795) is included at the bottom of each pattern group.

least-square fitting method to the observed XRD patterns, which are summarized in Table 1. All of the lattice constants, *a*, *b* and *c*, of Mn doped $Bi_4Ti_3O_{12}$ decreased with increasing Mn concentration, while *a* decreased and *b* and *c* increased as the Fe concentration increased. The volumes of both Mn- and Fe-doped $Bi_4Ti_3O_{12}$ unit cells decreased as the concentration of the dopant increased, though this volume reduction is much smaller in the Fe-doped ones. These results indicate that both Mn and Fe ions appear to be substituted at one of the cation sites, i.e., Bi or Ti sites.

Ionic radii for the cations in this system, Bi^{3+}, Mn^{2+}, Fe^{2+}, Mn^{3+}, Fe^{3+}, Ti^{4+}, Fe^{4+} and Mn^{4+}, with 6-coordination are 1.030, 0.830, 0.780, 0.645, 0.645, 0.605, 0.585 and 0.530 Å, respectively[5]. If Mn and Fe ions substitute at Bi sites, unit cell volume should decrease much since all the ionic radii

Table 1. Lattice parameters and volumes of $Bi_4(Ti_{1-x}M_x)_3O_{12}$ (M = Mn, Fe)

	x	a (Å)	b (Å)	c (Å)	V (Å^3)
pure	0	5.4496	5.4108	32.8167	967.66
	0.01	5.4428	5.4079	32.8073	965.65
Mn-doped	0.03	5.4378	5.4060	32.8062	964.40
	0.05	5.4362	5.4051	32.7956	963.64
	0.01	5.4456	5.4113	32.8298	967.42
Fe-doped	0.03	5.4433	5.4116	32.8296	967.06
	0.05	5.4402	5.4151	32.8272	967.06

with different valences are much smaller than that of Bi^{3+}. In the case of the substitution at Ti sites, only Fe^{4+} and Mn^{4+} can produce the experimentally observed reduction of volumes. However, the formation of an oxygen vacancy may appear for charge compensation in the crystal when trivalent Mn and Fe ions substitute at Ti^{4+} site, which may also lead to the reduction of the volume. Thus, an analysis of the doped crystal structure based on ionic substitutions alone cannot determine the substitution sites for Mn and Fe ions in $Bi_4Ti_3O_{12}$.

We employed the Mn- and Fe-$L_{2,3}$ XANES spectra to determine the valence state of doped Mn and Fe ions; these are in Fig. 3, shown alongside the spectra of Mn and Fe oxides. In general, $L_{2,3}$-edge XANES spectra of 3d transition metal is quite sensitive to change in valence state, since they can be obtained from the electric dipole transition from 2p to 3d states. In addition, profiles of $L_{2,3}$ XANES spectra can strongly reflect only nearest neighboring coordinations, since 3d orbital is a localized orbital. Then we are able to use the finger-print type analysis by comparing the spectrum of interest with standard oxides to determine the valence state of 3d transition elements. From the comparison of our observed Mn-$L_{2,3}$ spectra with those of the reference Mn oxides, we observe that the spectral fine structures of Mn ions in $Bi_4Ti_3O_{12}$ (Fig. 3(a)) show a similar profile to those of MnO_2, which indicates that the majority of Mn ions in $Bi_4Ti_3O_{12}$ exist as Mn^{4+}. On the other hand, the Fe-$L_{2,3}$ XANES profile of Fe ions in $Bi_4Ti_3O_{12}$ (Fig. 3(b)) is similar to that of α-Fe_2O_3, which suggests that most Fe ions in $Bi_4Ti_3O_{12}$ are in the 3+ state.

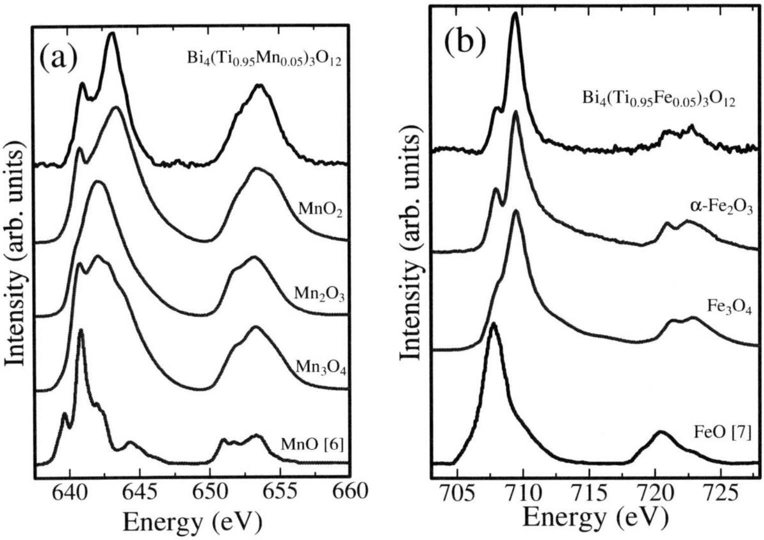

Figure 3. Observed $L_{2,3}$-edge XANES spectra for (a) Mn-doped and (b) Fe-doped $Bi_4Ti_3O_{12}$. The corresponding spectra of Mn and Fe oxides are also plotted for comparison.

Our observed Mn and Fe K-edge XANES spectra of Mn- and Fe-doped $Bi_4Ti_3O_{12}$ are in Fig. 4, alongside those of standard Mn and Fe oxides. The spectral fine structures of Mn and Fe in $Bi_4Ti_3O_{12}$ are different from any of the spectra of standard Mn and Fe oxides, which indicates Mn and Fe ions do not exist as simple oxides and the local environments of Mn and Fe ions are different from those in Mn and Fe oxides. This type of experimental comparison alone cannot determine the local environment of the doped ions in the case of K-edge XANES analysis, because K-edge XANES profile is a reflection of unoccupied p-states, which is rather delocalized than d-states. Then no standard spectra are available to be compared with the experimental spectrum of interest. Therefore, the additional information provided by theoretical calculations is required to resolve the final doped structure. These theoretical finger-prints can be compared with the observed XANES spectrum of interest to help identify the local environment of doped ions.

We performed these first-principles calculations to construct theoretical Mn-

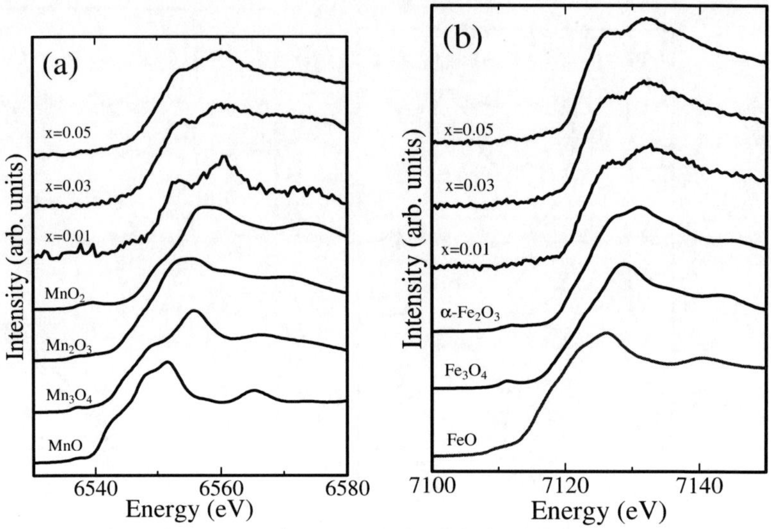

Figure 4. Observed K-edge XANES spectra for (a) Mn-doped and (b) Fe-doped $Bi_4Ti_3O_{12}$. The corresponding spectra of Mn and Fe oxides are also plotted for comparison.

K and Fe-K XANES spectra of Mn- and Fe-doped $Bi_4Ti_3O_{12}$, respectively. All the calculations were carried out using the augmented plane wave plus local orbital (APW+lo) solution of the Kohn-Sham equations, as implemented in the WIEN2k package [8]. The core-hole effect was directly introduced to fix the final state of the X-ray absorption process by removing one electron from the K-shell of Mn or Fe and placing an additional electron at the bottom of the unoccupied band, which is approximately correspond to the final states of Mn and Fe K-edge X-ray absorption process. We determined the energies associated with the calculated XANES spectra by taking the difference in total electronic energy of the initial (ground) and final (core-hole) states of the Mn and Fe K-edge X-ray absorption process.

Prior to the calculation of the Mn- and Fe-doped $Bi_4Ti_3O_{12}$, theoretical XANES spectra of MnO and FeO with a rock-salt structure were calculated using the 2 x 2 x 2 super-cell (64 atoms) of conventional unit-cell. The muffin-tin radii, R_{MT}s, of Mn, Fe, and O were set to 1.98, 1.98 and 1.30 a.u., respectively, and the plane wave cut-off K_{max} was 3.0 $Ry^{1/2}$. A 3 x 3 x 3 k-point mesh based on the Monkhorst-Pack scheme [9] was sampled in reciprocal space. The calculated spectra were broadened using a Gaussian function with width 1.16 and 1.25 eV for Mn and Fe K-edges, respectively. These peak broadenings correspond to the lifetime of a K core-hole [10]. We employed a generalized gradient approximation [11] for the exchange-correlation functional.

The calculated XANES spectra of MnO and FeO are compared to the experiments in Figs. 5(a) and (b), respectively. Characteristic features of XANES spectra of MnO and FeO were well reproduced by our calculations when the energy was corrected by 21.5 eV ($\Delta E/E = 0.3$ %) for MnO and 21.6 eV ($\Delta E/E = 0.3$ %) for FeO. It is noted here that the calculated spectra are sharper than the experimental ones. This is because the calculated spectra were broadened by taking only lifetime broadening effect but not taking the spectrometer broadening into considerations.

Next, we built super-cells for Mn- and Fe-doped $Bi_4Ti_3O_{12}$ by expanding the unit-cell of $Bi_4Ti_3O_{12}$ with a space group of I4/mmm along the b-axis twice, i.e., a 1 x 2 x 1 super-cell (76 atoms). From the results of Mn- and Fe-$L_{2,3}$ XANES analysis, valence states of Mn and Fe ions in $Bi_4Ti_3O_{12}$ have been determined to be 4+ and 3+, respectively. Then we constructed the models in which the Mn^{4+} and Fe^{3+} ions were substituted at Bi or Ti sites. There are two crystallographically independent sites for Bi (i.e., in a perovskite-type layer or a rock-salt-type layer) and two independent sites for Ti (i.e., Ti(2a) and Ti(4e)). We constructed eight types of models by replacing one of the Bi or Ti ions in one of these configurations by a Mn^{4+} or Fe^{3+} ion and by controlling the total number of electrons in the cell. When a Bi^{3+} ion is replaced by Mn^{4+} ion, for instance, one electron is removed from the cell at initial condition. Calculated spectral energies were shifted by the same amounts as for the case of MnO and FeO, i.e., 21.5 eV and 21.6 eV, respectively.

Calculated XANES profiles of Mn- and Fe-doped $Bi_4Ti_3O_{12}$ are compared with our experimental profiles in Figs. 6(a) and (b), respectively. In both figures, the calculated XANES spectra of Bi site models show significant differences compared to the observed profiles. On the other hand, both of the calculated spectra of the two types of the Ti site substitution models show better agreements to the experimental ones. It should be noted here that the spectral signatures of the Ti(4e) and Ti(2a) models are quite similar, which suggests the local environments of Mn ions are quite similar when they are substituted at Ti(2a) and Ti(4e) sites.

Unfortunately, the theoretical XANES finger-prints cannot determine which Ti site is preferable for the substitution of Mn and Fe ions in $Bi_4Ti_3O_{12}$. To determine the specific sites for substitution, we compared the total electronic energies calculated above within the density functional theory for the calcula-

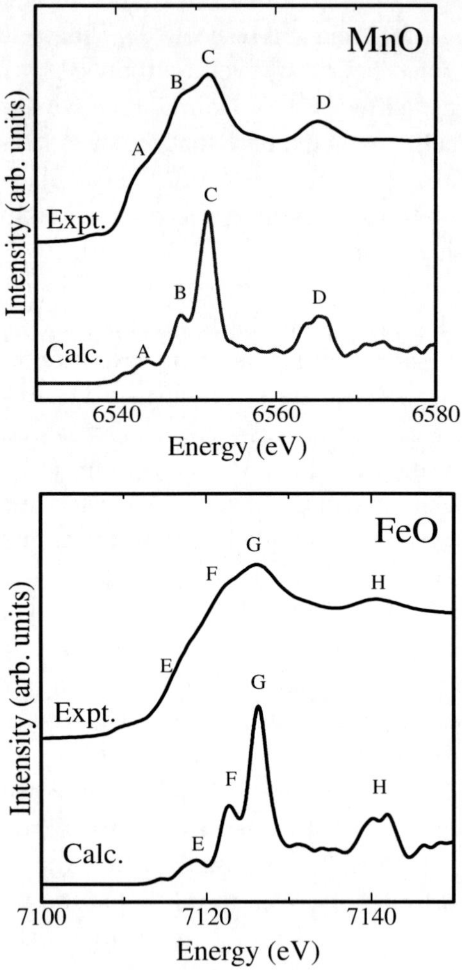

Figure 5. Comparison of K-edge XANES spectra between experiments and calculations of (a) MnO and (b) FeO.

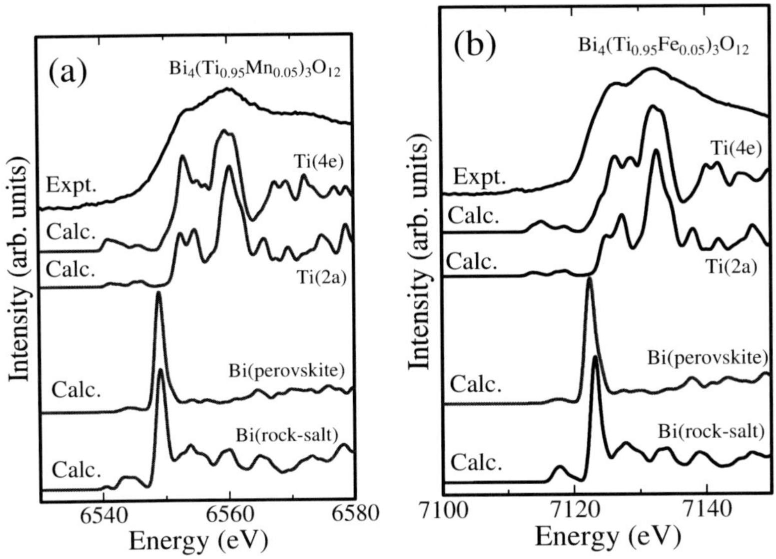

Figure 6. Comparisons of K-edge XANES spectra between experiments and calculations of (a) Mn-doped and (b) Fe-doped $Bi_4Ti_3O_{12}$.

tions of spectral energies. In the case of Mn substitution, the total electronic energy is lower by 0.16 eV/cell when Mn is substituted at Ti(2a) site than at Ti(4e), whereas that is lower by 0.30 eV/cell when Fe ion is at Ti(4e) site than at Ti(2a). This is an indication that Mn and Fe ions substitute at Ti(2a) and Ti(4e) sites, respectively.

5. Conclusion

In the current report, we have demonstrated the method for the local environment analysis of dilute magnetic elements in functional materials in an atomic scale by combined use of X-ray absorption near-edge structure (XANES) measurements and first-principles DFT calculations. Results for polycrystalline Mn- and Fe-doped $Bi_4Ti_3O_{12}$ fabricated by a solid state reaction method with different concentrations of Mn and Fe ions: x = 0.00, 0.01, 0.03, and 0.05 in $Bi_4(Ti_{1-x}Mn_x)_3O_{12}$ and $Bi_4(Ti_{1-x}Fe_x)_3O_{12}$, respectively, have been explained in detail as an example how to use our analytical method. After basic crystal structure analysis were carried out by the powder X-ray diffraction technique, substitution mechanism of Mn and Fe ions in $Bi_4Ti_3O_{12}$ was investigated by XANES analysis with the aid of the first-principles calculations. Mn- and Fe-$L_{2,3}$ XANES measurements determined the valence states of the Mn and Fe ions as 4+ and 3+, respectively. K-edge XANES analysis with the aid of the first-principles calculations revealed that Mn and Fe ions should substitute at Ti site. Finally, the first-principles total electronic energy calculations suggested that Mn and Fe ions should substitute at Ti(2a) and Ti(4e) sites, respectively. It has been confirmed here again that combined use of XANES measurements and first-principles DFT calculations is a powerful tool for the local environment analysis of the dilute magnetic elements in functional materials.

Acknowledgment

The authors would like to thank E. Shigemasa, H. Iwayama, E. Nakamura, and N. Kondo of UVSOR for their assistance with the Mn- and Fe-$L_{2,3}$ XANES measurements at BL4B in UVSOR, and T. Okajima of Saga-LS and K. Nitta of SPring-8 for their assistance with the Mn- and Fe-K XANES measurements at BL01B1 in SPring-8.

References

[1] T. Yamamoto, Y. Kawashima, Y. Kusakabe, S. Matsuda, Y. Mizuoka, Y. Nakade, and T. Okajima, *J. Phys.: Condens. Matter*, 21 (2009) 104211.

[2] I. Tanaka, T. Mizoguchi, M. Matsui, S. Yoshioka, H. Adachi, T. Yamamoto, T. Okajima, M. Umesaki, W. Y. Ching, Y. Inoue, M. Mizuno, H. Araki, and Y. Shirai, *Nature Mater.*, 2 (2003) 541.

[3] I. Tanaka, T. Mizoguchi, and T. Yamamoto, *J. Am. Ceram. Soc.*, 88 (2005) 2013.

[4] K. Nishimura, T. Yoshioka and T. Yamamoto, *IEEE Trans. Mag.*, 50 (2014) 2502306.

[5] R.D. Shannon, *Acta Cryst.*, A32 (1976) 751.

[6] B. Gilbert, B.H. Frazer, A. Belz, P.G. Conrad, K.H. Nealson, D. Haskel, J.C. Lang, G. Srajer, G. De Stasio, *J. Phys. Chem.*, A107 (2003) 2839.

[7] T.J. Regan, H. Ohldag, C. Stamm, F. Nolting, J. Luning, J. Stohr, R.L. White, *Phys. Rev.*, B64 (2001) 214422.

[8] P. Blaha, K. Schwarz, G. K. H. Madsen, D. Kvasnicka, and J. Luitz: WIEN2k, *An Augmented Plane Wave + Local Orbitals Program for Calculating Crystal Properties* (Karlheinz Schwarz-Technische Universitat Wien, Vienna, 2001).

[9] H. J. Monkhorst and J. D. Pack, *Phys. Rev.*, B 13 (1976) 5188.

[10] M. O. Krause and J. H. Oliver, *J. Phys. Chem. Ref. Data*, 8 (1979) 329.

[11] J. P. Perdew, K. Burke, and M. Ernzerhof, *Phys. Rev. Lett.*, 77 (1996) 3865.

In: Computational Materials ... ISBN: 978-1-63482-541-2
Editor: K. T. Kholmurodov, pp. 123-137

Chapter 12

THE SMALL NUCLEOTIDE CHAIN AND GOLD NANOPARTICLES INTERACTION INSIDE CARBON NANOTUBE: MOLECULAR DYNAMICS STUDY

Mirzoziz A. Khusenov[1,2,*]***, Ermuhammad B. Dushanov***[3]
and Kholmirzo T. Kholmurodov[2,3,4]
[1]Tajik Technical University named after M. S. Osimi,
Dushanbe, Republic of Tajikistan
[2]Moscow State University of Technology "STANKIN"
Moscow, Russia
[3]Joint Institute for Nuclear Research
Dubna, Moscow Region, Russia
[4]Dubna International University
Dubna, Moscow Region, Russia

Abstract

We simulated the interaction of the small nucleotide chain (NC) with gold nanoparticles (NPs) inside carbon nanotube (CNT) of an open ended boundary. Such system represents a great interest in many aspects of today biochemical and nanotechnological research (diagnostic aplications, drug delivery in cell, nanorobotic design and related manipulations). The

*E-mail address: mirzo85@inbox.ru

entire system (the NC chain, gold NPs and CNT) were allowed to interact with each other by the Van der Waals (VdW) forces only. The CNT was described through a quantum-chemistry potential, though the trajectory calculation for the whole NC–NP–CNT model was performed via the classical molecular dynamics (MD) approach. The Lennard-Jones short-ranged interaction is assumed between the NC, NP and CNT. We have carried out a series of MD simulations on different NC–NP–CNT configurations to investigate the pequliarities of NC–NP bonding and structural formation along with dynamical behavior inside a CNT matrix under rather a weak VdW interaction.

Keywords: Nucleotide Chain (NC), Gold Nanoparticles (NPs), Carbon Nanotube (CNT), Van der Waals (VdW) Interaction, Bond Formation, Molecular Dynamics (MD)

1. Introduction

In this work, the molecular dynamics (MD) simulations were performed on a small nucleotide chain (NC) (one purine and one pyrimidine) to investigate its interaction and binding processes with gold nanoparticles (NPs) that happen inside of a carbon nanotube (CNT). The small NC model interacting with a NP–CNT system one believe is an important stage in the understanding of the interaction mechanism of a full DNA or RNA molecule with NP and CNT. A linear sequence of nucleotides makes up the primary structure of DNA or RNA. The DNA and RNA primary structures consist of a linear sequence of nucleotides that are linked together by phosphodiester bonds. The understanding of the DNA interaction and binding mechanism with metallic nanoparticles and metallic surfaces embedded by the CNT environment represents a great interest in many aspects of today biochemical and nanotechnological research (diagnostic aplications, drug delivery in cell, nanorobotic design and related manipulations) [1-13].

Recent experimental and simulation studies involve the DNA interaction with highly localized proton beams or metallic NPs (such as Ag, Au, etc.), aimed on targeted cancer therapy through the injection of metal micro– or nanoparticles into the tumor tissue with consequent local microwave or laser heating. Along with DNA–NP also the DNA–CNT system represents a great interest in today biomedicine applications due to diagnostic and treatment of oncology diseases. Cancer, in which cells grow and divide abnormally, is one of the primary diseases with regards to how it responds to CNT drug delivery. Representing a revolutionarily potential for the biochemistry and medicine the use of CNTs in drug delivery has based on the enhancing of sufficient solubility and allowing of efficient tumor targeting. These aspects prevent CNTs from being cytotoxic and altering the function of immune cells. For today, cancer therapy involves surgery, radiation therapy, and chemotherapy. For example, recent experimental and simulation studies involve the interaction of DNA with highly localized high power beams and various nanoparticles (Ag, Au, etc.). These studies are aimed on targeted cancer therapy through the injection of metal micro- or nanoparticles into the tumor tissue with consequent local microwave or laser heating. Due to their good heat conductivities of NPs (Ag, Au, and so on) the experiments reveal that the only tumor cells to destroy, remaining normal cells undamaged. Nevertheless, such kind treatment methods are usually painful and kill normal cells in addition to producing adverse side effects [1-9].

CNTs as drug delivery vehicles have shown a potential interest due to a targeting of specific cancer cells with a lower dosage rather than conventional drugs have. With regard to the different aspects of the DNA interaction and carbon nanotubes (CNTs) there has also been a great discussion for today nano- and biotechnological innovation and application. For example, the use of the DNA–CNT system in DNA nanotechnology have intensively been discussed in molecular recognition processes, as a candidate material in cell drug delivery, as nucleic acid selection method (DNA aptamer) in SELEX (Systematic Evolution of Ligands by Exponential enrichment), so on. The conformational transition of aptamers (single chain DNA or RNA molecules that possess specific spatial structure) around CNT may cause some modification of the charge distribution on the CNT surface. It is worth noting that CNT surface has extremely sensitive to even a small change of the electrical charge of its environment. The replacement of even a single nucleotide for the DNA or RNA structure can modify, on the other hand, the charge environment around CNT. As a result, the CNT charge conductivity will be changed as well. So far, the DNA or RNA interactions with CNT could result to an essential modification for the charge distribution and consequent charge transfer through by the CNT surface. In the physics measurement the DNA–CNT charge distribution can be estimated trivially. This simple scheme from the point of view of application and diagnostic purposes has considered being one of the promising technologies in the DNA–CNT interaction processes with target proteins (say, of blood cells in human body) [5-13].

So far, the studying molecular mechanism of the NC–NP–CNT reaction has seen strongly been motivated which MD (molecular dynamics) simulations were under the intense research in recent years.

2. Model and Method

We have simulated a three component molecular system consisting of a two nucleotide chain (NC), gold nanoparticles (NPs) and a carbon nanotube (CNT) at different temperature conditions (T=100, 200 and 300 K) (Fig. 1). Two nucleotides (one pyrimidine and one purine) as in the primary structure of DNA (Figs. 2 and 3) were relaxed in the vicinity of gold particles. The nucleotide chain was located from gold atoms at the distances $[5 \div 10\ \mathring{A}]$, i.e. within a range of Van der Waals (VdW) forces (Figs. 4).

The entire system (the nucleotide chain, gold atoms and carbon nanotube) were allowed to interact with each other via the VdW potential only. For describing of VdW interactions we used Lennard-Jones (LJ) potential, which is commonly in use for simulation of liquids and condensed phases. The LJ potential look like:

$$U = \sum_{ij} \left[4\varepsilon_{ij} \left(\left(\frac{\sigma_{ij}}{r_{ij}} \right)^{12} - \left(\frac{\sigma_{ij}}{r_{ij}} \right)^{6} \right) \right] \tag{2.1}$$

Here $\sigma_{ij}, \varepsilon_{ij}$ are LJ interaction parameters for ith and jth atoms; r_{ij} is interatomic distance; q_i, q_j are partial atomic charges. The cross-section in-

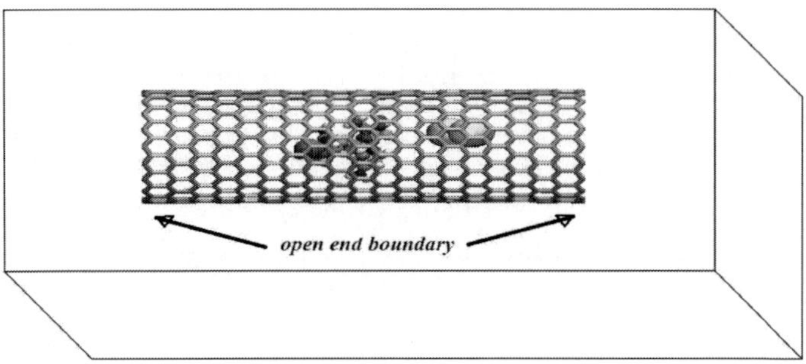

Figure 1. Schematic view of the simulation system.

teraction parameters were defined using the Lorentz-Berthelot mixing rule: $\sigma_{ij} = (\sigma_i + \sigma_j)/2, \varepsilon_{ij} = (\varepsilon_i \varepsilon_j)^{1/2}$

A classical molecular dynamics study was performed using the DL_POLY_2.20 [14–15] general-purpose code. The NVT ensemble with a Berendsen thermostat and a Verlet leapfrog scheme were employed. The integration time step of the dynamical equations of motion was 1 fs. The force field parameters for the NC (nucleotide chain) and CNT (carbon nanotube) molecules were choosen from the DL_FIELD database [15]. The intramolecular interactions for the NC chain were described using the LJ, combined with angular and dihedral bonding potentials. The LJ, cross-interaction and bond parameters for the NC–NP-CNT system were the same as in our paper [13].

For the describing CNT we have employed the quantum chemistry Tersoff potential, so far a co-called hybrid approach (the combination of quantum chemistry potential and classical trajectory calculations) has been realized to investigate the NC interaction with with CNT. It is worth noting that the Tersoff potential is efficiently used for MD simulation of systems that contain carbon, silica, germanium, etc. alloys. In CNTs we have chemical bonding is hybridization sp2 (as graphite), which is stronger than sp3 bond (of diamond). The nature of chemical bonding in CNTs is described by quantum chemistry, through the process of orbital hybridization. The Tersoff potential in hybrid MD simulations correctly describes the nature of covalent bonding in carbon nanotube; it allows the breaking and formation of chemical bonds, that is associated with hybridization process. Tersoff potential is pair wise potential, but coefficient in attractive term depends on local environment, thus, Tersoff potential possesses a many body nature (see, [13] and references therein).

$$U_{ij} = f_C[f_R(r_{ij}) - \gamma_{ij} f_A(r_{ij})] \tag{2.2}$$

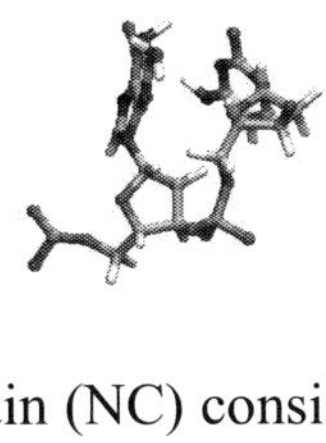

Figure 2. A two nucleotides chain (NC) consisting of one pyrimidine (C, cytosine) and one purine (G, guanine).

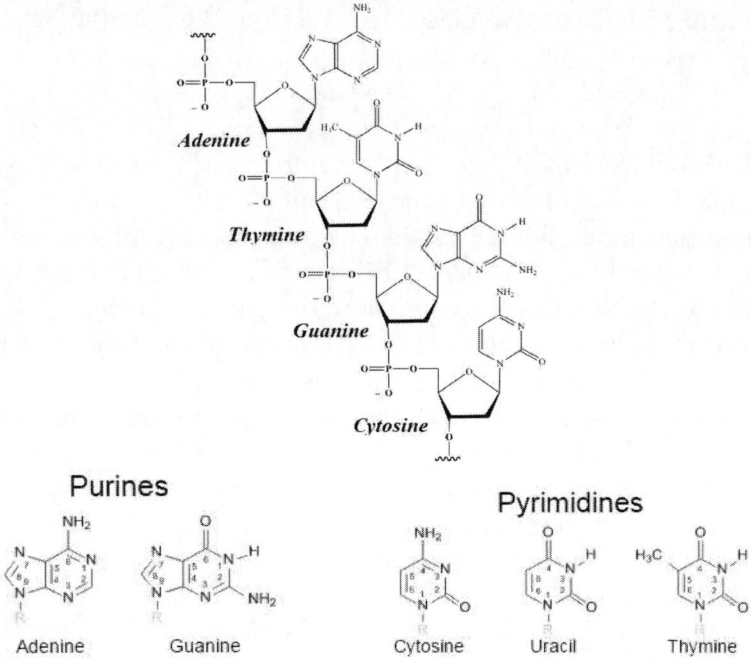

Figure 3. (Top) A linear sequence of nucleotides that make up the primary structure of DNA or RNA. A primary DNA or RNA structure consists of a linear sequence of nucleotides that are linked together by phosphodiester bonds. (Bottom) Nucleotides consist of 3 components: (1)Nitrogenous base – A (Adenine), G (Guanine), C(Cytosine), T (Thymine, present in DNA only) and U (Uracil, present in RNA only); (2) 5-carbon sugar which is called deoxyribose (found in DNA) and ribose (found in RNA); (3) One or more phosphate groups.

3. Results and Discussion

The MD simulations were realized in the temperature range $T = 100$, 200 and 300 K. For the NC–NP–CNT system we simulated 1, 2 and 3 gold particles as shown in Figs. 4. First, we have generated three positionally slightly different gold particles enable to interact with the NC as NC–(w)Au, NC–(s)Au and NC–(i)Au. Here the indexes (w),(s) and (i) imply the formation of preferrably weak, strong and intermediate bonds between the NC–NPs, respectively. Next, we considered the different possible combinations of a two gold particle and a nu-

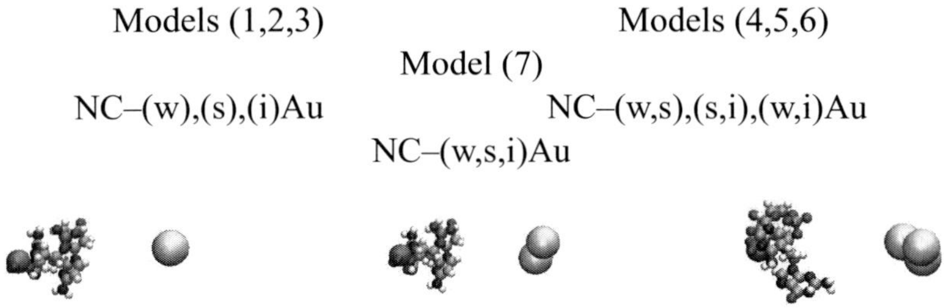

Figure 4. The models of the nucleotide chain (NC) interacting with 1, 2 and 3 gold nanoparticles (NPs). The positions of one of two phosphorus (P) atoms of the NC and gold NPs are shown respectively by the dark brown and yellow spheres.

cleotide chain as NC–(w,s)Au, NC–(s,i)Au and NC–(w,i)Au. Finally, we make a small triple gold cluster and simulated the NC–NPs system as NC–(w,s,i)Au. So far, we have investigated in total seven molecular configurations for the NC–NPs–CNT under the same simulation and temperature conditions:

(1) NC–(w)Au, (2) NC–(s)Au, (3) NC–(i)Au,
(4) NC–(w,s)Au, (5) NC–(s,i)Au,
(6) NC–(w,i)Au, (7) NC–(w,s,i)Au

The total potential energy of the NC–NPs–CNT system at the temperatures T=100, 200 and 300 K is shown in Fig. 5. The energy profiles, as seen from Fig. 5, converge smoothly to their equilibrium values, so the the system equilibration has satisfactory controlled.

In Figs. 6-8 we present the MD simulation snapshots and distance calculations for the one gold particle system as NC–(w)Au, (s)Au, (i)Au, two gold particle NC–(w,s)Au, (s,i)Au, (w,i)Au and three gold particle NC–(w,s,i)Au systems, respectively. The results indicate on possibilities of formation of different type contacts (bonding) for the nucleotides during their collective interaction with gold atoms. Under the similar condition and being relaxed inside the same CNT matrix, the NC–NP interaction may produce a different scenario of a bond formation. We thus observe a concurrency between the NC intramolecular internal vibrations and a weak VdW interaction with gold NPs. The NC chain can form with some gold atoms a close contact, while with other the resultant bonding looks rather week.

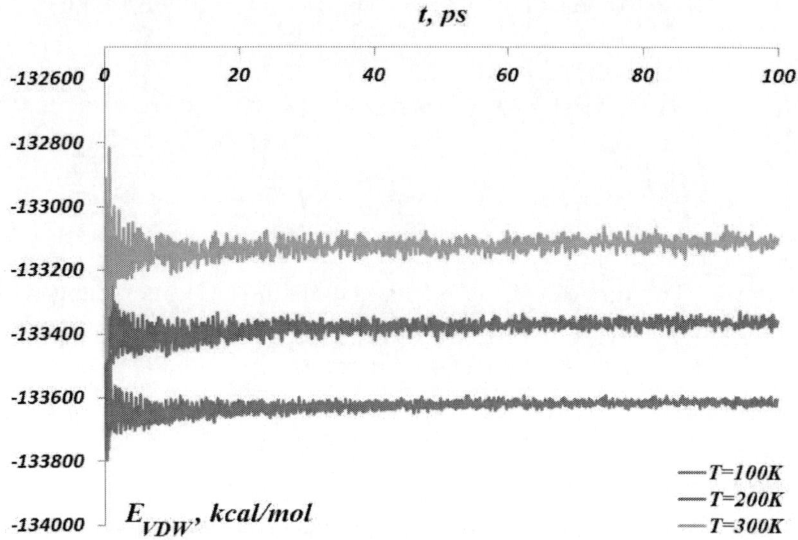

Figure 5. The total potential energy of the NC–NPs–CNT system at the temperatures T=100, 200 and 300 K.

In Figs. 9-11 the calculation results are presented for dihedral energy of NC at temperature T=100 K. The MD results are compared between the one gold particle system as NC–(w),(s),(i)Au, two gold particle NC–(w,s),(s,i),(w,i)Au and three gold particle NC–(w,s,i)Au systems, respectively. From Figs. 9, for a single gold case, we can observe that while NC–NP contact (bonding) is weak (say, for the models NC–(w)Au and NC–(i)Au) at the initial stages of dynamics the oscillations of the dihedral energy are also high for. For the NC–(s)Au model, where formation of a relatively strond bond is preferrable, the dihedral energy vibrations look rather smooth. From Figs. 10-11 for the two and three gold particle systems (the models 4-6 and 7) the NC–NPs close contacts are always preferrable. In other words, the effect of two and three gold atomic clustering and consequent collective motions has seen to generate stronger interaction with NC.

The MD simulations have been completed and compared for the NC–NPs–CNT system at low and high temperatures (Figs. 12–14). In Figs. 12 we present the NC(P)–(w),(s),(i)Au interaction distances, the NC dihedral and angular energies for the temperatures T=100, 200 and 300 K. The above observations, with regard to possibilities of preferrably weak, strong and intermediate bond

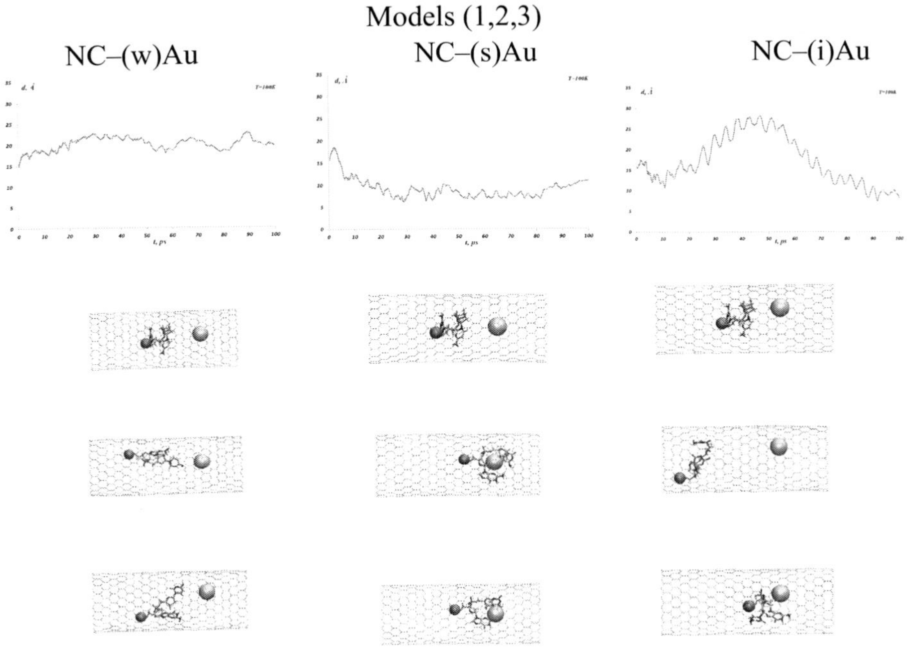

Figure 6. The interaction distances NC(P)–(w)Au, NC(P)–(s)Au and NC(P)–(i)Au (from left to right) for the moments of t=0, 50 and 100 ps (from top to bottom).

formations between the NP–NC, in part remain valid. However, at higher temperatures as the NC intramolecular oscillations grown the pisture of the NC–NPs bonding inside the CNT chain essentially modifies. We observe more fluctuations in the NP–NC bonding processes, not only for single gold atomic cases (Figs. 12) but also for the two (Figs. 13) and three (Figs. 14) particles. The comparison of the MD results shown in Figs. 12-14 are straighforward.

4. Concluding Remarks

In this work, molecular dynamics (MD) simulations have been performed to study the peculiarities of an interaction mechanism and a bond formation for a small nucleotide chain (NC) with gold nanoparticles (NPs) that happen in the presence of a carbon nanotube (CNT) matrix. The NC structure conisiting of

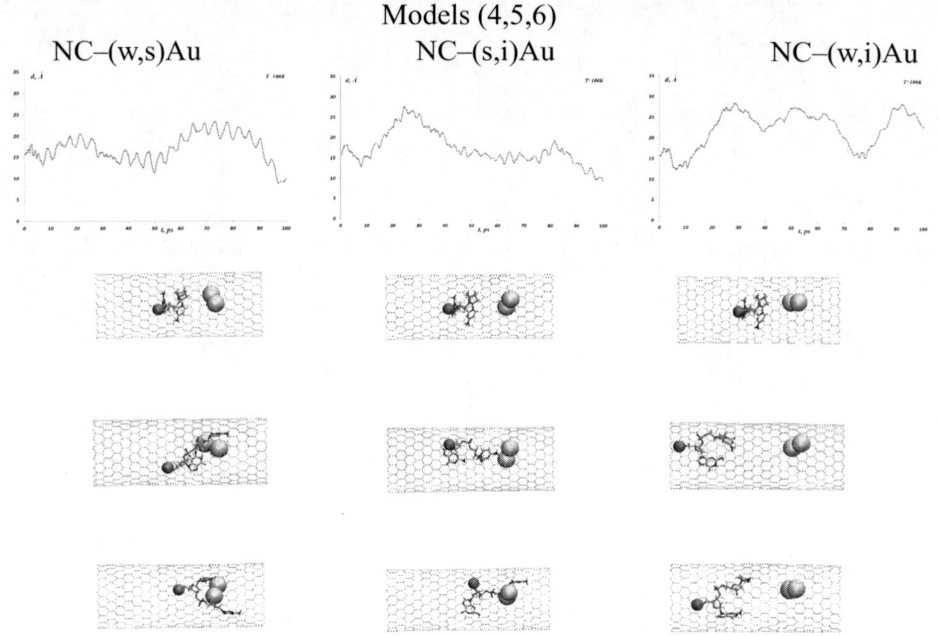

Figure 7. The interaction distances NC(P)–(w,s)Au, NC(P)–(s,i)Au and NC(P)–(w,i)Au (from left to right) for the moments of t=0, 50 and 100 ps (from top to bottom).

one purine (guanine) and one pyrimidine (cytosine) was brought in contact with gold NPs inside of a CNT chain of an open ended boundaries. We assumed the existing of the only Van der Waals forces in the system and the Lennard-Jones short-ranged interaction between the NC, NP and CNT was used. Nevertheless, the CNT was described through a quantum-chemistry potential, so far, a so-salled hybrid approach combining the classical trajectory calculation and quantum chemistry potential was realized.

We have carried out a series of MD simulations with different NC–NPs–CNT molecular models to investigate the pequliarities of VdW interactions, bond and structural formations along with the dynamical behavior. The NC intermolecular motions were estimated from MD data thereby building the the distance distributions, the angular and dihedral (torsional) bond energy graphs vs simulation time.

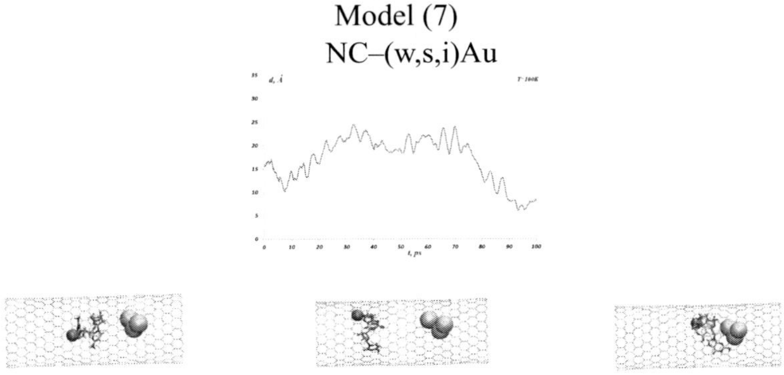

Figure 8. The interaction distance NC(P)–(w,s,i)Au for the moments of t=0, 50 and 100 ps (from top to bottom).

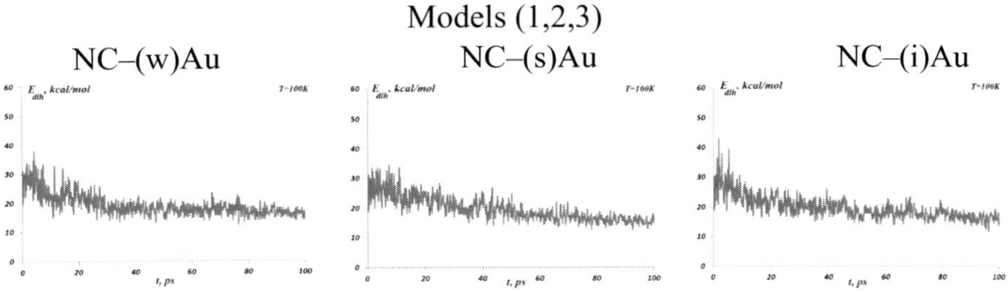

Figure 9. The dihedral bond energies vs time for the NC interacting with (w)Au, (s)Au and (i)Au (from left to right).

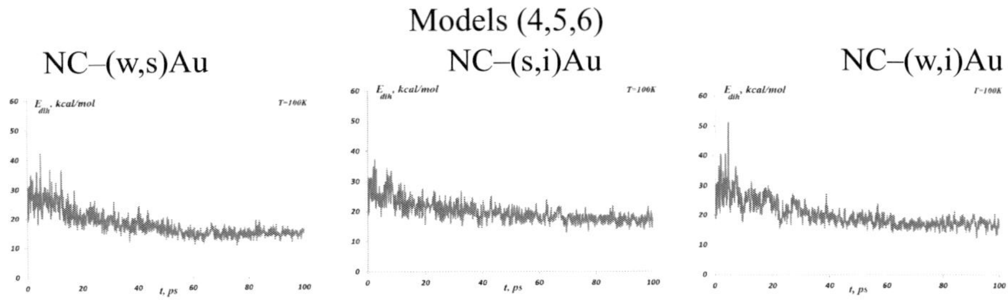

Figure 10. The dihedral bond energies vs time for the NC interacting with (w,s)Au, (s,i)Au and (w,i)Au (from left to right).

Model (7)

NC–(w,s,i)Au

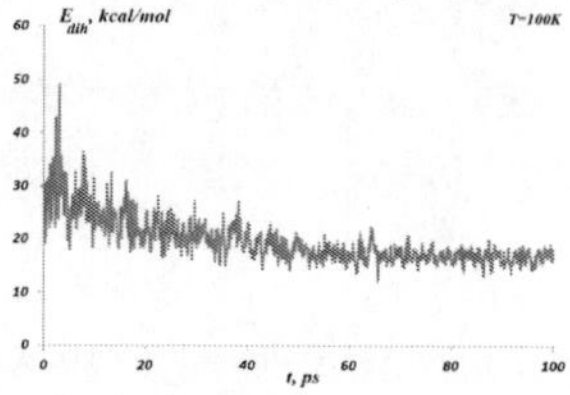

Figure 11. The dihedral bond energy vs time for the NC interacting with (w,s,i)Au.

Models (1,2,3)

NC–(w)Au NC–(s)Au NC–(i)Au

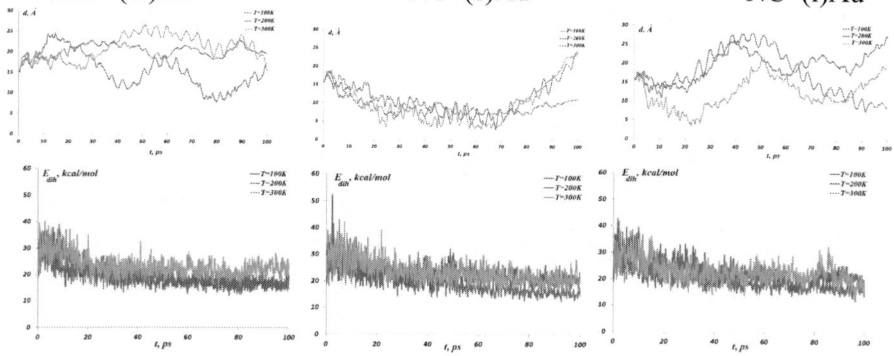

Figure 12. The NC(P)–(w)Au, NC(P)–(s)Au and NC(P)–(i)Au interaction distances, the NC dihedral energies at the temperatures T=100, 200 and 300 K (from left to right)

The MD simulation results have shown that depending on the relative NC–NPs positions a different scenario of a bond formation might be possible. We have fixed up the three types of the NC–NPs binding (viz., weak, strong and intermediate) as a result of the collective NC intermolecular motion and weak VdW NC–NPs interaction that realized inside of a CNT matrix. The snapshots of the NC-NPs interaction and positional changes inside of a CNT demonstrate the binding processes at different temperatures from T=100 to 300 K. The NC intermolecular vibrations and weak NC–NPs VdW interactions are a kind of concurrent effect defining the NC–NPs structure pequliarities within a CNT.

In conclusion, the small NC–NP–CNT model one believe is an important

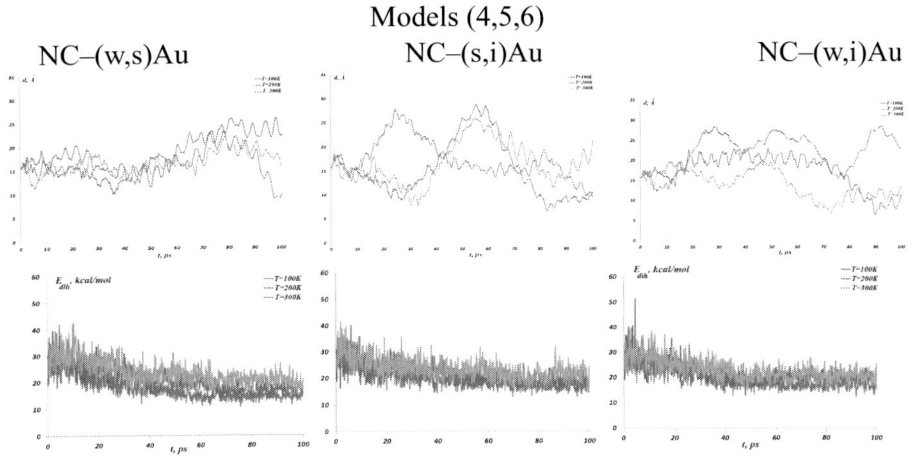

Figure 13. The NC(P)–(w,s)Au, NC(P)–(s,i)Au and NC(P)–(s,i)Au interaction distances, the NC dihedral energies at the temperatures T=100, 200 and 300 K (from left to right).

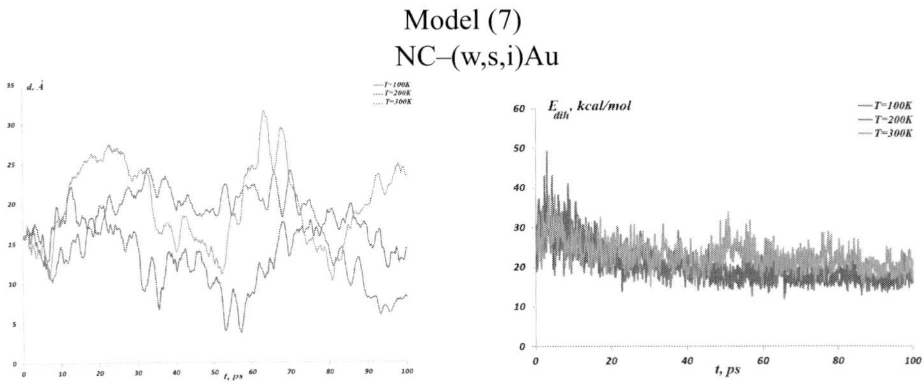

Figure 14. The NC(P)–(w,s,i)Au interaction distance and the NC dihedral energy at the temperatures T=100, 200 and 300 K.

stage in the understanding of the interaction mechanism of a full DNA or RNA molecule with NPs and CNT. The DNA–NPs-CNT system represents a great interest in many aspects of today biochemical and nanotechnological research. In this regard, one firstly have to mention the DNA–CNT devices for the diagnostic aplications, the chemical or drug delivery in cell, related nanorobotic design and so on.

References

[1] R. Dunford, A. Salinaro, L. Cai, N. Serpone, S. Horikoshi, H. Hidaka, J. Knowland, Chemical oxidation and DNA damage catalysed by inorganic sunscreen ingredients, *FEBS Letters*, 418(1), 1997, 87-90. http://dx.doi.org/10.1016/S0014-5793(97)01356-2.

[2] Monti S., Carravetta V., Zhang W., and Yang J., ”Effects Due to Interadsorbate Interactions on the Dipeptide/TiO2 Surface Binding Mechanism Investigated by Molecular Dynamics Simulations”, *J. Phys. Chem. C*, 111, 2007, 7765-7771.

[3] SantaLucia J.(Jr.), ”A unified view of polymer, dumbbell, and oligonucleotide DNA nearest-neighbor thermodynamics”, *Biochemistry*, 95, 1998, 14601465.

[4] Tulpan D., Andronescu M., Leger S., ”Free energy estimation of short DNA duplex hybridizations”, *BMC Bioinformatics*, 2010, 11:105.

[5] SantaLucia J. (Jr.) and Hicks D., ”The thermodynamics of DNA structural motifs”, *Annu. Rev. Biophys. Biomol. Struct*, 2004, 415440.

[6] Breslauert K.J., Franks R., Blockers H., and Markyt L.A., ”Predicting DNA duplex stability from the base sequence”, *Biochemistry*, 83, 1986, 3746-3750.

[7] Freyre-Fonseca V., Delgado-Buenrostro N.L., Gutirrez-Cirlos E.B., Caldern-Torres C.M., Cabellos-Avelar T., Snchez-Prez Y., Pinzn E., Torres I., Molina-Jijn E., Zazueta C., Pedraza-Chaverri J., Garca-Cullar C.M., Chirino Y.I., ”Titanium dioxide nanoparticles impair lung mitochondrial function”, *Toxicology Letters*, 202, 2011, 111119.

[8] Trouiller B., Reliene R., Westbrook A., Solaimani P., and Schiestl R.H., ”Titanium Dioxide Nanoparticles Induce DNA Damage and Genetic Instability In vivo in Mice”, *Cancer Res*, 69(22), 2009.

[9] Saquib Q., Al-Khedhairy A.A., Siddiqui M.A., Abou-Tarboush F.M., Azam A., Musarrat J., "Titanium dioxide nanoparticles induced cytotoxicity, oxidative stress and DNA damage in human amnion epithelial (WISH) cells", *Toxicology in Vitro*, 26, 2012, 351361.

[10] Dunford R., Salinaro A., Cai L., Serpone N., Horikoshi S., Hidaka H., Knowland J., "Chemical oxidation and DNA damage catalysed by inorganic sunscreen ingredients", *FEBS Letters*, 418, 1997, 87-90.

[11] T.A. Hilder, J.M. Hill, Carbon nanotubes as drug delivery nanocapsules, *Current Applied Physics*, 8(3-4), 2008, pp. 258261. http://dx.doi.org/10.1016/j.cap.2007.10.011.

[12] Kh. Kholmurodov (Ed.), "Molecular Dynamics of Nanobistructures", *Nova Science Publishers Ltd.*, 2011, 210p., ISBN: 978-1-61324-320-6.

[13] Khusenov, M., Dushanov, E. and Kholmurodov, K. "Molecular Dynamics Simulations of the DNA-CNT Interaction Process: Hybrid Quantum Chemistry Potential and Classical Trajectory Approach". *Journal of Modern Physics*, 5, 2014, 137-144. doi: 10.4236/jmp.2014.54023.

[14] Forester T.R. and Smith W. "Dl_POLY_2.0: A general-purpose parallel molecular dynamics simulation package". *Journal of Molecular Graphics*, 14(3), 1996, 136–141.

[15] Yong C.W. "Dl_FIELD — a force field and model development tool for DL_POLY". In: Richard Blake, editor, *CSE Frontiers. STFCs Computational Science and Engineering Department (CSED)*. Science and Technology Facilities Council, STFC Daresbury Laboratory, 2010, 38–40.

In: Computational Materials ... ISBN: 978-1-63482-541-2
Editor: K. T. Kholmurodov, pp. 139-150© 2015 Nova Science Publishers, Inc.

Chapter 13

LIPID ENVIRONMENT PROMOTES DIMERIZATION OF TRANSMEMBRANE HELICES

Andrey S. Kuznetsov[1*]***, Anton A. Polyansky***[1,2]***,***
Pavel E. Volynsky[1]***, Roman G. Efremov***[1,3]
[1]M. M. Shemyakin and Yu. A. Ovchinnikov
Institute of Bioorganic Chemistry,
Russian Academy of Sciences, Moscow, Russia.
[2]Department of Structural and Computational Biology,
Max F. Perutz Laboratories,
University of Vienna, Vienna, Austria
[3]Higher School of Economics, Moscow, Russia

Abstract

Membrane receptors are among the most important molecules in the cell. Their functioning determines the response of the cell to the external conditions. Also, they provide cellular communication and recognition mechanisms. So, their proper functioning is a vital question of life. There is a lot of data today about these receptors. Many three-dimensional (3D) structures are known, genetic causes of some diseases are determined and many therapeutic agents acting on the membrane proteins are developed. But today with all this knowledge we cannot exactly predict the effects

---

*Email address: andrej.kuznecov@phystech.edu

of new mutations in these proteins on their activity and cannot decipher the mechanisms of their action in detail. On the other side, cell membrane consists of phospholipids acting as a matrix for the proteins and other molecules like glycolipids. Of course, exact lipid composition of the membrane affects the proteins behavior, but this phenomenon is reversible. Many papers describe how single-point mutations act on the protein activity, another large area of study includes determination of the influence from the lipid bilayer properties. But little is said about how the protein itself induces local perturbations on the membrane. Taking this into account, we decided to examine the simplest case of the membrane protein with a help of atomistic computer simulations of molecular dynamics. Using transmembrane (TM) helical peptide representing TM-domain of human glycophorin A in hydrated lipid bilayer, we showed that it can induce local perturbations of lipid properties, and the characteristic size of the pattern is 10-15 Å. So, the peptides can feel each other on large distances (about 20 Å), and this estimation is confirmed by calculation of the association free energy. We determined that the interaction interface for dimer formation also serves as a binding site for lipids. They are removed from this site during dimerization process, but stay nearby in the dimer structure filling the gaps next to the contact of helices. We assume that native sequences generate specific distributions of lipids around them to find each other in the membrane and have specific surface relief to strengthen the interaction in case of dimerizing proteins by lipid binding to the formed dimer.

Keywords: Free Energy Calculations, Transmembrane Helix, Lipid Membrane, Lipid Properties Distribution

1. Introduction

Integral membrane proteins provide cell signaling and membrane transport functions for the cell. All these proteins are associated with the lipid bilayer with via their transmembrane (TM) domains that often contain α-helical segments. These helices do not stay isolated in the lipid environment – instead, they tend to form helical pairs (hairpins) or more complex aggregates (helical bundles). Such an assembly is prerequisite to get a protein in its active state. Improper packing of helices can result in severe diseases, like myeloma, urinary bladder cancer or other [1, 2]. Knowing the details of helices interaction in membrane on molecular level, one will be able to regulate biological functioning of these systems in a goal-oriented manner. The clue to the functioning of many membrane receptors (like receptor tyrosine kinases) is the association of their TM α-helices in membrane. As membrane is not a homogeneous medium [3], one can consider that lipids affect protein dimerization and therefore modulate their activity.

In most cases, membrane proteins have large extra- and intracellular domains and also the membrane composition varies depending on cell type. That is why the native system is very difficult to study with both theoretical and experimental techniques. Often, the model systems containing two TM peptides in hydrated lipid bilayer or membrane mimetic are utilized to simplify the task. Nowadays, considerable progress has been achieved in determination of 3D structures of TM helical dimers. Some tens of spatial structures of dimers have been solved by NMR in different membrane mimics – detergent micelles and lipid/detergent bicelles [4-7]. Also, computer molecular modeling techniques have proved to be efficient in prediction of 3D structures of the helix-helix complexes [8, 9]. Based on this data, a hypothesis was putted forward that dimerization of helices is determined by special amino acid patterns in the sequence – the so-called called dimerization motifs [10, 11]. The most famous one is the glycophorin-like motif: small-xxx-small, where small indicates residues with small side chain (Gly, Ala, rarely – Ser), while x denotes any residue. In addition, the important property of a dimer is its affinity, free energy of dimerization (ddG) that can be estimated using both experimental and computational techniques [12-15]. It is very sensitive to protein-protein interaction between monomers and to interactions between protein and lipid bilayer as well [16, 17]. The fact that the calculated Potential of Mean Force profiles (PMF-profiles) for TM peptides interaction reach zero only at the distances larger than 20 Å [18] provokes a question: how do the helices initially feel one another? If lipid bilayer has a heterogeneous structure itself and shows a predisposition for lipid clusters formation [3, 19], can it serve as a medium for broadcasting membrane proteins interactions? To answer this question, we performed an analysis of the dynamic properties of GpA TM domain monomer and dimer in hydrated lipid bilayer using atomistic molecular dynamics (MD) simulations. Main accent in this work was done on exhaustive monitoring of structural and dynamic behavior of lipids near the peptide surface. To evaluate a putative role of such environmental effects in dimerization, single TM helices were simulated in lipid bilayers as well.

2. Materials and Methods

2.1. Model Peptides

The calculations were performed for a peptide representing GpA TM domain (SEPEITLIIFGVMAGVIGTILLISYGIRR). All-atom starting structure of monomer were built in ideal α-helical conformation using PyMOL program. Arg and Glu residues, along with N- and C- termini were taken in their charged forms. Dimeric structure was taken from the Protein Data Bank (PDB id: 1AFO, [20]).

2.2. Construction of Systems, MD Protocol

Construction of the starting system for simulation was done by inserting the peptide monomer or dimer into a model lipid bilayer comprised of 128 molecules of palmitoyloleoylphosphatidylcholine (POPC). It was done by removal of some (10 to 20) lipid molecules to form a pore for protein using genbox utility supplied with the GROMACS software. Water molecules (SPC model) were added to fill all the cavities in the simulation box following by removal of waters located inside the bilayer to prevent pore formation.
All simulations were performed using GROMACS 4.5.6 package for molecular dynamics and 43a2x parameter set from the GROMOS96 force field [21]. Simulations were carried out with a time step of 2 fs and imposed 3D periodic boundary conditions. A twin-range 10/12 Å spherical cutoff function was used to truncate van-der-Waals interactions. Electrostatic interactions were treated using the particle-mesh Ewald summation (real space cutoff 10 Å and 1.2 Å grid with fourth-order spline interpolation). The isothermal-isobaric (NPT) ensemble was used with a semi-isotropic pressure of 1 bar and a constant temperature of 315 K. The pressure and the temperature were controlled using the V-rescale thermostat and Berendsen barostat with 1.0 and 0.1 ps relaxation parameters, respectively, and a compressibility of $4.5 \times 10^{-5}\ bar^{-1}$ for the barostat. Protein, lipids and water molecules were coupled separately.

In the beginning of the simulations, the systems were equilibrated by energy relaxation via 50000 iterations of the steepest descent minimization followed by heating from 5 K to 315 K during 50 ps MD run and 50 ns MD run at 315 K with fixed positions of the protein atoms. Such equilibration is necessary for minimization of the bilayer distortion induced by insertion of the peptide. Finally, 100 ns collection MD runs were carried out. POPC bilayer without a peptide also was also subjected to all the simulations and analysis as a reference system. The stability of the peptides was checked by calculating geometrical parameters of a dimer and analyzing the secondary structure of monomers. Two independent MD trajectories were obtained for each system.

2.3. Calculation of Free Energy Profiles

Potential energy profile was calculated as a function of helix-helix distance using umbrella sampling with integration of a mean force [22]. The distance between centers of mass was considered as a reaction coordinate. In total, 32 windows were taken with distance varying from 7 to 22.5 Å. A set of reference structures was obtained by translating monomers along the line connecting their centers of mass. Initial structure for each window was generated using 50 ns MD equilibration when the monomers positions were fitted to the target reference structure with a given distance between them. The 50 ns production MD runs were carried out in each window with a harmonic force constant of $1000\, kJ\, mol^{-1}\, nm^{-2}$. Error estimation was done by dividing MD traces into parts, which were used for independent profiles calculations. The calculation was repeated twice.

2.4. Lipid Properties Analysis

Lipid density was mapped on a plane (thus giving 2D maps) for 5 separate layers across a normal to the bilayer. They were namely, polar heads (2 layers), main part of acyl chains (2 layers) and central part consisting tails of acyl chains. All the maps were centered on peptide and averaged over multiple frames of MD. Periodic boundary conditions were taken into account. Lipid density was calculated using orthogonal grid. Average positions of peptide's Cα atoms were also shown on these maps. Cylindrical projections were made using same protocol but in cylindrical coordinate system using alpha-helix axis as reference. 3D distribution of lipid acyl chains density was also examined.

3. Results and Discussion

3.1. Molecular Dynamics and Free Energy Calculations

Monomer and dimer demonstrate no significant changes in secondary structure during unrestrained MD simulations in POPC. Only terminal charged residues

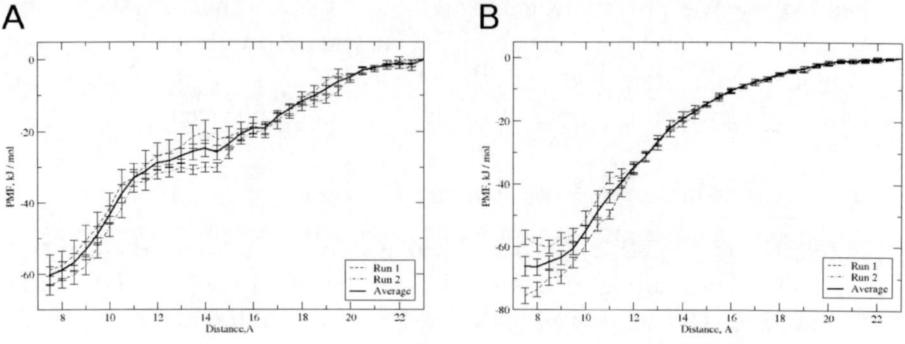

Figure 1. Free energy of association for glycophorin A in POPC bilayer. Total energy profiles (A) and energy of direct protein-protein interaction (B).

have unfolded during 50 ns of MD and their conformations in different trajectories were not similar. As proposed, geometrical parameters of the dimer were well preserved. This suggests that the single TM segments of GpA are stable and their structure is independent on the monomer-dimer transition, that is in agreement with simulations done before by other groups [22]. In the course of MD runs, root-mean-square deviation (RMSD) for the backbone from the starting structure was about 1.2 Å. The peptides demonstrate very strong tendency for dimerization, the depth of energy minimum slightly differs depending on the starting conditions. Wild-type (WT) form of GpA has an energy gain of approximately $60 \pm 3\, kJ/mol$ (Figure 1, A). This is in a good agreement with experimental and theoretical estimations [22, 23]. It is clear that the monomers are interacting on the distances larger than 15 Å, where there is no direct protein-protein contact (Figure 1, B). The only factor that can be responsible for such a long-term interaction can be the membrane. So, lipid properties were analyzed in different ways to determine how can lipids play a role in association of TM helices.

3.2. Membrane Properties Around Peptides

As mentioned above, lipids play an important role in process of GpA TM segments dimerization, and such a possibility has been discussed earlier [24]. When the density maps for pure POPC bilayer were compared with ones containing peptides, a significant difference was observed (Figure 2, A). Whereas

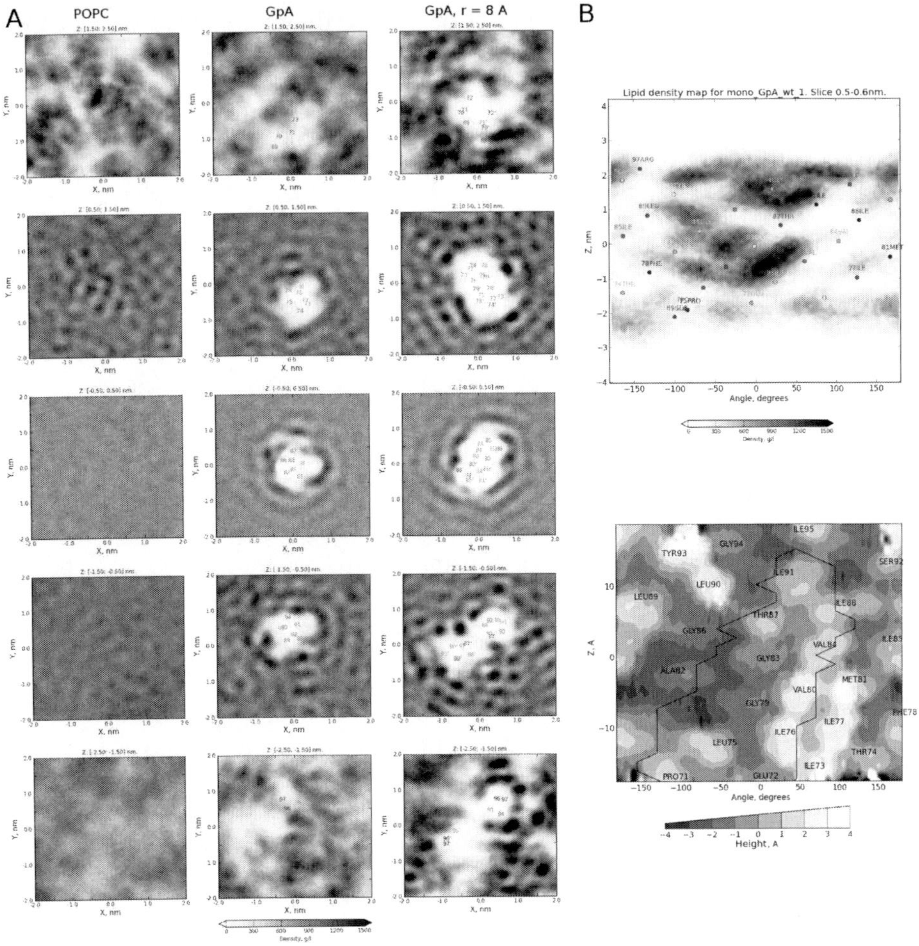

Figure 2. Lipid density maps for different parts of the membrane for pure POPC bilayer and bilayer containing monomer or dimer of GpA (A). Five rows of panels correspond to the following slices along the membrane normal (from up to bottom): polar heads; main part of acyl chains; central part consisting tails of acyl chains; main part of acyl chains; polar heads. Cylindrical map of the surface relief (bottom) compared to lipid density map (top) for TM peptide of GpA. A contour (thin black line) shows the helix-helix dimerization interface (B).

polar heads form mosaic picture even in pure bilayer, probably, corresponding to lipid clusters [3], acyl chains in that case are very flexible and give almost uniform distribution of density on timescales longer than 5-10 ns. When any peptide (monomer or dimer) is present in a bilayer, it becomes surrounded by tightly bound lipids. The number of these lipids and their distribution depends on the nature of TM sequence, and results in non-uniform distribution of the lipid density. The effect is almost the same as observed lipid clustering around integrins [25], but it is detected in more simple system. The main result is the formation of strongly frozen lipid region on one side of monomeric TM helix corresponding to dimerization interface (Figure 2, B) of GpA helix, which represents a hydrophilic hollow surrounded by longer hydrophobic side chains. It was surprising that lipids bind to this region. Another interesting thing is that lipids do not lose their flexibility upon this binding. No significant changes were found in the mean lipid order, so not just a single lipid molecule is involved in this interaction, but two or three acyl chains replace each other on a nanosecond timescale. The positions of density maximums remain the same because of geometrical parameters of the surface that suits the acyl chain size followed by formation of such a solvation shell around peptide in a membrane plane. We therefore assume that surrounding lipids can broadcast some influences of the peptide to a large neighboring area of the bilayer and serve as a markers for proteins to find each other and orient properly in the membrane.

When two peptides come closer, their lipid patterns become overlapped at the distance c.a. 18-20 Å between centers of masses of the monomers. There is no direct protein-protein contact at that stage, but the lipid perturbations start to evolve. New density maxima appear as monomers approach each other, and the interfacial maximum vanishes at the separation about 10-12 Å. Formation of the dimer is accompanied by appearance of many additional lipid density peaks on the sides of the structure (Figure 2, A). The composition of the interhelical region may play an important role allowing lipid acyl chains to fill the cavity regions and stabilize the dimer structure.

Conclusion

Transmembrane (TM) domains of the receptors play crucial role in functioning of the cell membrane, but one should pay attention to the lipids too. They not only affect the dimerization/oligomerization ability of each other via structural adaptation upon tight packing, but can directly interact with binding sites on the protein surface. It was shown that even single helices induce perturbations in the bilayer and the formation of asymmetric complex between TM domain of proteins and surrounding lipids may have significant effect at the initial stages of proteins interaction. Also, lipids acyl chains may serve as molecular glue filling all the voids or hollows on the surface of the membrane proteins and strengthen the dimerization. These interactions can be highly specific in case of large integral membrane proteins, but they take place in case of single α-helices too. Even in case of such a strong dimer as glycophorin A lipids may play an important role in process of its association.

Acknowledgments

This work was supported by the Russian Scientific Foundation (grant 14-14-00871 awarded to R.G.E.). Authors express their gratitude to the Russian Foundation for Basic Research (grant 13-04-00825) and RAS Programmes Molecular and Cellular Biology and Basic fundamental research for nanotechnologies and nanomaterials. Access to computational facilities of the Joint Supercomputer Center of RAS (Moscow) is greatly appreciated.

References

[1] Bergsagel P.L., Chesi M., Molecular classification and risk stratification of myeloma, *Hematol. Oncol.*, 31 (2013) 38-41.

[2] Al Hussain T.O., Akhtar M., Molecular basis of urinary bladder cancer, *Adv. Anat. Pathol.*, 20 (2013) 53-60.

[3] Pyrkova D.V., Tarasova N.K., Krylov N.A., Nolde D.E., Pentkovsky V.M., Efremov R.G., Dynamic clustering of lipids in hydrated two-component membranes: results of computer modeling and putative biological impact, *J. Biomol. Struct. Dyn.*, 31 (2013) 87-95.

[4] Mineev K.S., Bocharov E.V., Volynsky P.E., Goncharuk M.V., Tkach E.N., Ermolyuk Y.S., Schulga A.A., Chupin V.V., Maslennikov I.V., Efremov R.G., Arseniev A.S., Dimeric structure of the transmembrane domain of glycophorin a in lipidic and detergent environments, *Acta Naturae*, 3 (2011) 90-98.

[5] Bocharov E.V., Mayzel M.L., Volynsky P.E., Mineev K.S., Tkach E.N., Ermolyuk Y.S., Schulga A.A., Efremov R.G., Arseniev A.S., Left-handed dimer of EphA2 transmembrane domain: helix packing diversity among receptor tyrosine kinases, *Biophys. J.*, 98 (2010) 881-889.

[6] Bocharov E.V., Mineev K.S., Goncharuk M.V., Arseniev A.S., Structural and thermodynamic insight into the process of weak dimerization of the ErbB4 transmembrane domain by solution NMR, *Biochim. Biophys. Acta*, 1818 (2012) 2158-2170.

[7] Bocharov E.V., Lesovoy D.M., Goncharuk S.A., Goncharuk M.V., Hristova K., Arseniev A.S., Structure of FGFR3 transmembrane domain dimer: implications for signaling and human pathologies, *Structure*, 21 (2103) 2087-2093.

[8] Polyansky A.A., Chugunov A.O., Volynsky P.E., Krylov N.A., Nolde D.E., Efremov R.G., PREDDIMER: a web server for prediction of transmembrane helical dimers, *Bioinformatics*, 30 (2014) 889-890.

[9] Polyansky A.A., Volynsky P.E., Efremov R.G., Multistate organization of transmembrane helical protein dimers governed by the host membrane, *J. Am. Chem. Soc.*, 134 (2012) 14390-14400.

[10] Melnyk R.A., Kim S., Curran A.R., Engelman D.M., Bowie J.U., Deber C.M., The affinity of GXXXG motifs in transmembrane helix-helix interactions is modulated by long-range communication, *J. Biol. Chem.*, 279 (2004) 16591-16597.

[11] Doura A.K., Kobus F.J., Dubrovsky L., Hibbard E., Fleming K.G., Sequence context modulates the stability of a GxxxG-mediated transmembrane helix-helix dimer, *J. Mol. Biol.*, 341 (2004) 991-998.

[12] Duneau J.P., Vegh A.P., Sturgis J.N., A dimerization hierarchy in the transmembrane domains of the HER receptor family, *Biochemistry*, 46 (2007) 2010-2019.

[13] Russ W.P., Engelman D.M., TOXCAT: A measure of transmembrane helix association in a biological membrane, *Proc. Natl. Acad. Sci. U. S. A.*, 96 (1999) 863-868.

[14] Cymer F., Sanders C.R., Schneider D., Analyzing oligomerization of individual transmembrane helices and of entire membrane proteins in E. coli: a hitchhikers guide to GALLEX, *in Protein Supersecondary Structures*, Kister A. E. ed., Humana Press, 2013, pp. 259-276.

[15] Polyansky A.A., Volynsky P.E., Efremov R.G., Structural, dynamic, and functional aspects of helix association in membranes: a computational view, *Adv. Protein Chem. Struct. Biol.*, 83 (2011) 129-161.

[16] Smith S.O., Eilers M., Song D., Crocker E., Ying W., Groesbeek M., Metz G., Ziliox M., Aimoto S., Implications of threonine hydrogen bonding in the glycophorin A transmembrane helix dimer, *Biophys. J.*, 82 (2002) 2476-2486.

[17] Caputo G.A., Analyzing the effects of hydrophobic mismatch on transmembrane α-helices using tryptophan fluorescence spectroscopy, *in Membrane Proteins*, Ghirlanda G., Senes A. ed., Humana Press, 2013, pp. 95-116.

[18] Volynsky P.E., Polyansky A.A., Fakhrutdinova G.N., Bocharov E.V., Efremov R.G., Role of dimerization efficiency of transmembrane domains in activation of fibroblast growth factor receptor 3, *J. Am. Chem. Soc.*, 135 (2013) 8105-8108.

[19] Parton D.L., Tek A., Baaden M., Sansom M.S.P., Formation of raft-Like assemblies within clusters of influenza hemagglutinin observed by MD simulations, *PLoS Comput. Biol.*, 9 (2013) e1003034.

[20] MacKenzie K.R., Prestegard J.H., Engelman D.M., A transmembrane helix dimer: structure and implications, *Science*, 276 (1997) 131-133.

[21] Van Der Spoel D., Lindahl E., Hess B., Groenhof G., Mark A. E., Berendsen H.J.C., GROMACS: Fast, flexible, and free, *J. Comput. Chem.*, 26 (2005) 1701-1718.

[22] Hnin J., Pohorille A., Chipot C., Insights into the Recognition and Association of Transmembrane α-Helices. The free energy of α-helix dimerization in glycophorin A, *J. Am. Chem. Soc.*, 127 (2005) 8478-8484.

[23] Hong H., Chang Y.C., Bowie J. U., Measuring transmembrane helix interaction strengths in lipid bilayers using steric trapping, *in Membrane Proteins*, Ghirlanda G., Senes A. Ed., Humana Press, 2013, pp. 37-56.

[24] Janosi L., Prakash A., Doxastakis M., Lipid-modulated sequence-specific association of glycophorin A in membranes, *Biophys. J.*, 99 (2010) 284-292.

[25] Mehrbod M., Mofrad M.R.K., Localized lipid packing of transmembrane domains impedes integrin clustering, *PLoS Comput. Biol.*, 9 (2013) e1002948.

In: Computational Materials ... ISBN: 978-1-63482-541-2
Editor: K. T. Kholmurodov, pp. 151-163

Chapter 14

PROTEIN–PROTEIN INTERACTION IN PHOTOSYNTHETIC ELECTRON TRANSFER CHAIN: COMPUTER SIMULATION

Ilya B. Kovalenko[1,2,*]***, Anna M. Abaturova***[1]***,***
Aleksandra N. Diakonova[1]***, Dmitry M. Ustinin***[1]***,***
Sergei S. Khruschev[1]***, G.alina Yu. Riznichenko***[1]
and Andrei B. Rubin[1]
[1]Biological Faculty,
Lomonosov Moscow State University,
Leninskye Gory, Moscow, Russia
[2]Federal Research and Clinical Center of the FMBA of Russia,
Moscow, Russia

Abstract

We develop a method of multiparticle Brownian dynamics and apply it to computer simulation of interactions of electron transport proteins. This method makes it possible to study the effects of various factors on protein-protein complex formation. Some of them are connected with electrostatic interactions (ionic strength and pH of the solution, charge-changing point protein mutations). Other factors are molecule concentration, viscosity of the solution, size and shape of the reaction volume. We will show how Brownian dynamics takes into account these factors by

[*]E-mail address: ikovalenko78@gmail.com

the example of interaction of photosynthetic electron transport proteins plastocyanin and cytochrome f.

Keywords: Brownian Dynamics, Computer Simulation, Protein, Photosynthesis, Electrostatic Interaction

1. Introduction

Plastocyanin (Pc), a small copper protein, makes transient complexes with cytochrome *f* (Cyt *f*) of cytochrome *bf* complex and with the luminal part of photosystem I shuttling electrons from cytochrome *bf* complex to P700, the reaction center of PSI [8] in the luminal space of chloroplasts.

The process of interaction of an electron carrier protein with a multienzyme complex is assumed to proceed in five steps: (1) approach of a small protein to the complex due to diffusion and electrostatic attraction; (2) interaction between separate parts of the molecules determined mainly by electric charge configuration, spatial orientation of a mobile molecule coupled to the formation of the initial predocking complex; (3) final binding and formation of the active complex capable of electron transfer; (4) actual electron transfer; and (5) complex dissociation. To a good approximation, diffusion and geometrical shapes of the interacting protein molecules as well as long-range electrostatic interactions play the key role in the initial preliminary complex formed via the rigid-body association [9]. In their turn, molecular flexibility in the protein-protein interface and hydrophobic interactions of proteins may be important in the transformation of the initial complex into the final complex.

Brownian dynamics (BD) models are usually used to simulate protein-protein interaction kinetics. In BD simulations the interaction of two individual proteins is investigated in solution, the proteins are treated as rigid or semi-rigid bodies, their geometric shape is considered at the atomic resolution, and electrostatic interactions of the proteins are presented in great details. There are BD models for the association of a pair of individual molecules Pc and Cyt *f* [20, 21, 24]. There are some approaches which can be characterized as a multiparticle BD models. For example, the SDA program can be used to simulate the diffusion of multiple proteins, in dilute or concentrated solutions [7]; the BDBOX program can simulate either single molecules or multicomponent systems of diverse, interacting molecular species using flexible, coarse-grained bead models [3]; the Brownmove - a software for particle based reaction diffusion dynamics in crowded cellular environments [25].

Recently, we have developed a new BD-based method for multiparticle 3D computer simulation of simultaneous diffusion and interactions of several hundreds of proteins. Our approach makes it possible to simulate reaction kinetics and obtain reaction time courses directly from the computer simulation. This method was used to study Pc–Cyt *f* complex formation for wild type (wt) and mutant Pc at a variety of ionic strength values in solution [11] and in the chloroplast thylakoid lumen [12]. The method was further developed to take into account the electrostatic interactions of proteins with the thylakoid membrane in the thylakoid lumen [10]. The method was applied for studying of electrostatic interactions effect on the rates of complex formation between PSI and mobile carrier proteins Pc and Fd at various ionic strength values to explain the experimentally observed non-monotonic dependences of the association second-order rate constant on the ionic strength for Pc–PSI and Fd–PSI reactions [14]. This approach was also used to study kinetic characteristics of Fd and FNR complex formation in solution [15]. Complex formation of Pc with PSI and Cyt f was simulated in the case of cyanobacteria [16, 17]. Besides, this method was applied for prediction of binding areas and possible complex structures for several pairs of test proteins [13].

2. The Computational Method

Each protein molecule in the computer model is represented by a 3D rigid body. The model scene is a parallelepiped with mirror boundary conditions. Protein molecules can be mobile or fixed at their initial positions. All molecules are considered to be mobile when modelling reaction in solution, membrane proteins are fixed along two opposite sides of the parallelepiped when the scene represents the chloroplast thylakoid lumen. In the process of simulation the mobile protein molecules diffuse in the reaction volume under the action of the random Brownian force and electrostatic force. As shown in [4], the Langevin equation can be used for the mathematical description of such process. The Langevin equation for translational motion has the following form:

$$\xi_{\text{tr}}^{x} \frac{dx}{dt} = f_x(t) + F_x,$$

where x is a coordinate, ξ_{tr}^{x} is a viscous friction coefficient at x, $f_x(t)$ and F_x are projections of the random Brownian force and electrostatic force onto the x axis, respectively. The random Brownian force has a normal distribution with zero mean value and dispersion $2kT\xi_{\rm tr}^{x}/\triangle t$. Here k is the Boltzmann constant, T is temperature, $\triangle t$ is time step. The projection of electrostatic force $F_x = -q\,d\varphi/dx$, here q is the charge, φ is the electrostatic potential.

The Langevin equation for rotational motion:

$$\xi_{\rm rot}^{x}\frac{d\alpha}{dt} = m_x(t) + M_x,$$

where α is the angle of rotation, $\xi_{\rm rot}^{x}$ is the viscous friction coefficient for rotational motion around x axis, $m_x(t)$ and M_x are the moments of the random Brownian force and electrostatic force relative to the axis x, respectively. In hydrodynamic calculations protein molecules were presented as prolate ellipsoids with larger axis coinciding with axis of molecule minimum inertia and having the same moment of inertia. Viscous friction coefficients for ellipsoids were calculated using formulae from [1]. Dynamic viscosity of medium was considered as equal to that of water ($1 \cdot 10^{-3}$ kg m^{-1} s^{-1} at 293 K). For the description of each object position we have chosen the Cartesian coordinate system with mirror boundary conditions at the edges (the scene coordinate system). Every object is assigned to its local Cartesian coordinate system, which coincides with the respective spheroid axes. Random numbers were produced by the LEcuyer random number generator [19]. Time step for numerical integration of Langevin equation (100 ps) was chosen so that average displacement of molecules during one step was about the size of electrostatic potential grid (2 Å).

Electrostatic interactions play an important role in protein-protein complex formation. Partial charges of the proteins produce heterogeneous electrostatic field. The electrostatic potential rapidly decreases with the distance from the protein surface due to screening by polar water molecules and ions of salt in solution. When a protein is far from other proteins, it moves under the action of the Brownian force but not the electrostatic force, and hence its motion is determined only by the Brownian diffusion. As the protein comes closer to other proteins, its motion is governed by the electrostatic field of the latter so that a favorable position for the initial complex formation can be reached. Our calculations show that for Pc and Cyt *f* proteins the distance where electrostatic interactions start to influence the protein-protein complex formation rate is 28 Å. In the simulation the electrostatic interactions are taken into consideration only when the distance between two protein surfaces is less than 35 Å because electrostatic interactions are very weak at greater distances. The magnitude of electrostatic field at these and greater distances from the protein surface is considered to be zero. The formal charges on the proteins were calculated as described in [5]. The pKa values of the amino acid residues were taken from [2]. The solution pH is taken into account in the Henderson-Hasselbalch equation to calculate the charges on amino acids atoms, for all of the runs pH is 7.

To a good approximation, the encounter complexes are formed by rigid-body association [9]. In the model, when two protein molecules come closer to each other within the docking distance during the diffusion process, the proteins are considered to form an initial encounter complex. However, molecular flexibility in the protein-protein interface and hydrophobic interactions of proteins may be important for the transformation of the initial encounter complex into the final complex. Our method does not explicitly simulate a transformation of the initial encounter complex into the final complex. Instead, this transformation is implicitly taken into account by introducing the probability p of final complex formation. The details are given in [11].

By calculating directly the number of complexes formed at each time step and plotting it versus time, we obtain a complex formation kinetic curve. We can compare obtained kinetic curve with the kinetic curves which are registered experimentally, or we can calculate rate constants of corresponding biochemical reaction from obtained kinetic curve. To do this, we consider a simple kinetic scheme of a second-order reaction

$$A + B \xrightarrow{k} AB,$$

where k is the second-order association rate constant, A denotes the first protein, B denotes the second protein. Bimolecular kinetic scheme matches the calculated kinetic curves well, if the solution concentration is not very high.

3. Results and Discussion

The Brownian Dynamics method makes it possible to study the effects of various factors on protein-protein complex formation. Some of them are connected with electrostatic interactions, they are ionic strength and pH of the solution, charge-changing point protein mutations. Proteins from different species also have different electrostatic field. Other factors are molecule concentration, viscosity of the solution, size and shape of the reaction volume. We will show how Brownian dynamics takes into account these factors by the example of interaction of proteins plastocyanin and cytochrome f.

3.1. Electrostatic Potential Field of Pc and Cyt *f*

Fig. 1 displays surface electrostatic potential. Although both proteins have negative net charge (−2 for Cyt *f* and −8 for Pc at pH 7), the binding area on Cyt *f* is positively charged, while the binding area on Pc is negatively charged, which results in molecules attraction and proper orientation and promotes the complex formation.

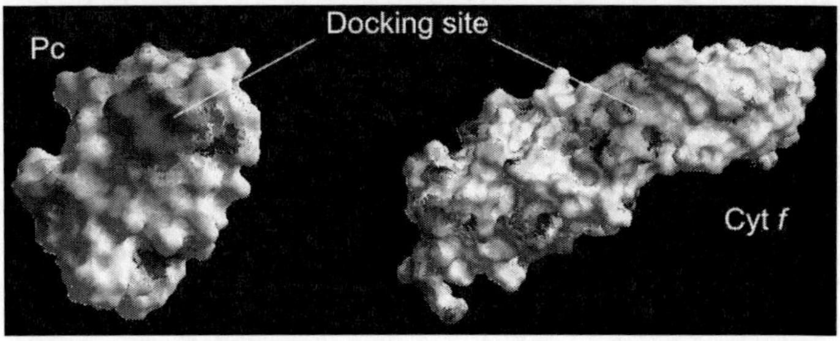

Figure 1. Surface electrostatic potential of Pc and Cyt *f*. The potential value at every point is shown by blue (positive potential) and red (negative potential) color.

3.2. Second Order Rate Constant Dependence on Ionic Strength

Attraction between the negatively charged regions of Pc and the positive ones of Cyt *f* assists the proteins to form a proper complex. However protein electrostatic field is screened by polar water molecules and ions in the solution. Increasing ionic strength leads to a decrease in proteins electrostatic interaction due to a protein electric charge screening by salt ions. If the rate constant decreases monotonically with increasing ionic strength, it means that electrostatic interactions between protein favor their binding. Brownian dynamics can successfully simulate the dependence of the second order rate constant of protein-protein interaction upon ionic strength. Such calculated dependencies for Pc–Cyt *f* reaction for different Pc mutants are presented in Fig. 2 together with experimental curves [11].

There is strong correlation with the experimental results. For the wild type plastocyanin the model value of the rate constant is in the error interval of the experimental value. For the mutant Q88E the rate constant is higher than for the wild-type plastocyanin both for the experiment and the model, though the model

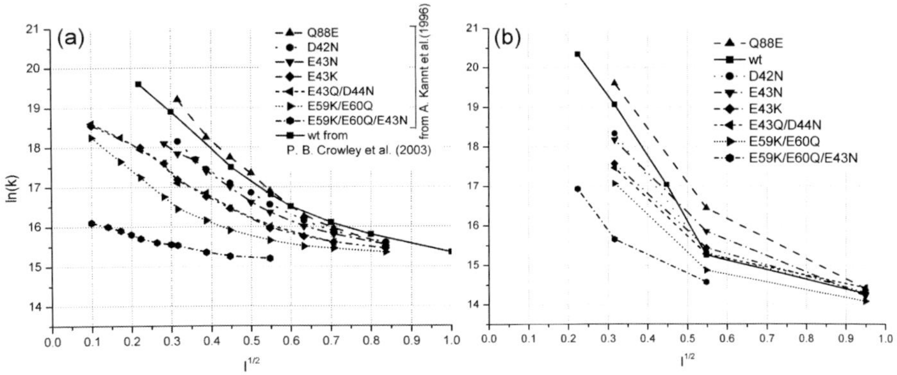

Figure 2. Dependence of the second-order rate constant of Pc-Cyt *f* association on the square root of ionic strength for different Pc mutants (Reproduced from [11] by permission of IOP Publishing. All rights reserved.).

value is higher than experimental. Q88E mutation adds a negative charge to the binding site at tyrosine 83 where a large region of negative potential is found which leads to the stronger attraction of this negative patch on Pc to the positive patch on cyt *f* and to the increase in the rate constant. In all other mutants the net charge of plastocyanin becomes more positive and this leads to the decrease in the reaction rate.

Brownian dynamics model successfully simulates the effect of pH. Fig. 3 shows experimental and calculated dependence of protein-protein complex formation rate on pH. Both experimental and calculated curves are bell-shaped with a wide plateau from pH 5.5 to 9. As pH value in the luminal space of thylakoids may change from 7 up to 5 in the light, the constant rate of the Pc–Cyt *f* complex formation in wide pH range makes it possible to maintain the high rate of electron transfer via Pc.

Using Brownian dynamics, we simulated the process of complex formation between Pc and Cyt *f* in the thylakoid lumen. The dependence of the rate of Pc-Cyt *f* association in the thylakoid on lumen width (distance between membranes) was assessed by varying the latter from 200 to 7 nm keeping constant the number of molecules (270 each) and the membrane area (322 x 322 nm). The Cyt *f* molecules were membrane-bound and exposed in the lumen. The results of numerical simulation [12] displayed in Fig. 4 show that the association rate is highest at a lumen width of 8 nm; it declines at larger widths because of the

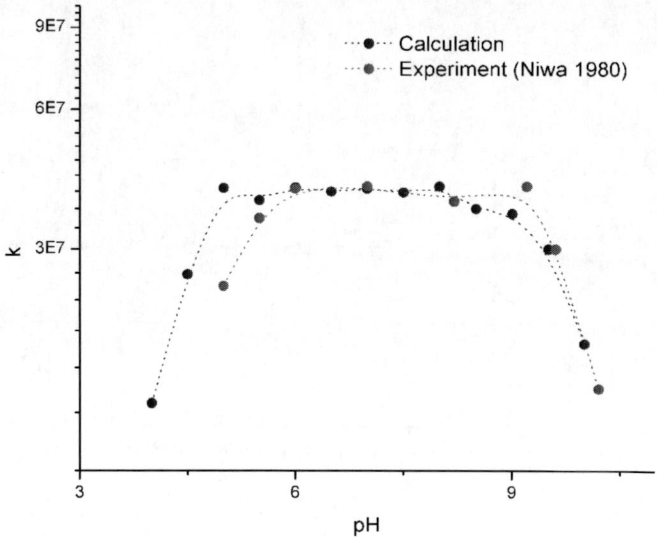

Figure 3. Calculated and experimental [18] dependencies of second-order rate of Pc–Cytf complex formation on pH.

decrease in effective Pc concentration, but is also lower at smaller widths. At z < 7 nm the rate constant abruptly drops to zero because Cyt f with attached Pc protrude from the membrane up to 7 nm, so if the opposite membrane is closer than this, the complex cannot be formed.

A photosynthetic membrane surface charge is another factor affecting the rate of Pc–Cyt *f* complex formation in thylakoid lumen: its negative electrostatic field prevents the negatively charged Pc from approaching the membrane-bound Cyt *f*, so the more negative is the membrane charge, the less is the rate constant of complex formation. Calculated second-order rate constants for Pc–Cyt *f* complex formation at different membrane charge densities σ and different ionic strength values are presented in Fig. 5 [10]. The most negative charge density of a membrane $\sigma = -42$ mC/m^2 in the simulation corresponds to a membrane consisting only of lipids (about 17% of membrane area is occupied by charged lipids).

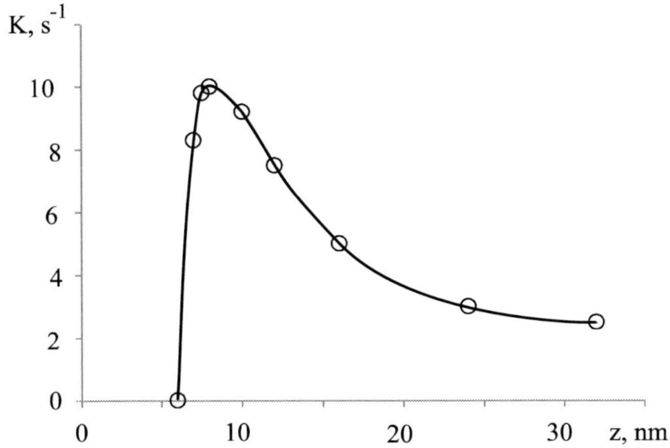

Figure 4. The rate of Pc-Cyt *f* complex formation normalized to Cyt *f* concentration in the thylakoid versus the intermembrane distance z at a constant number of molecules.

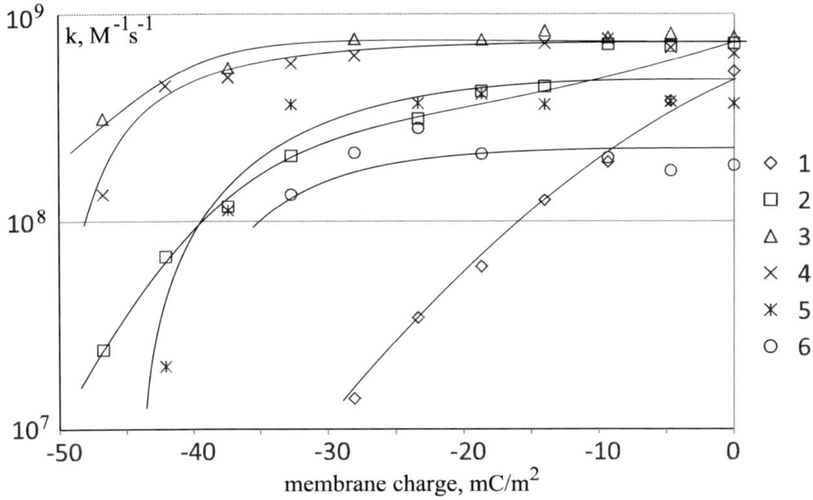

Figure 5. Dependences of the Pc–Cyt f binding rate constant on membrane charge density at ionic strengths of: 1, 0; 2, 10; 3, 30; 4, 50; 5, 100; 6, 150 mM. Modeling space 100x100x10 nm, pH 7.0; 25 protein molecules of each type.

4. Conclusion

The simulation method presented in this paper can be applied for diffusion and functioning studying of many macromolecules that interact in the heterogeneous interior of subcellular systems. A characteristic feature and novelty of our method gives an opportunity to use it for studying of several protein molecules interaction simultaneously making it possible to simulate the formation of a large number of complexes taking place in solution or cell compartments and to monitor this process real-time kinetics.

The method allows to study the role of electrostatic interactions in protein complex formation. Though it does not clearly consider conformational changes and electron transport, it can help to reveal the effect of electrostatic potential changes upon the kinetic parameters of protein interaction and to evaluate the rate constant values of complex formation.

Acknowledgments

This work was supported by the grants of Russian Foundation for Basic Research 15-07-08927, 14-04-00302 and 15-04-08681. Calculations were performed using Lomonosov Moscow State University supercomputers "Lomonosov" and "Chebyshev".

References

[1] V.A. Bloomfield. Survey of biomolecular hydrodynamics. On-Line Biophysics Textbook: *Separations and Hydrodynamics*, 2000.

[2] R.M.C. Dawson, D.C. Elliott, W.H. Elliott, K.M. Jones. *Data for biochemical research*. Oxford Science Publications, OUP, Oxford, 1986.

[3] M. Dlugosz, P. Zielinski, J. Trylska. Brownian dynamics simulations on CPU and GPU with BDBOX. *J. Comput. Chem.*, 32 (2011), 2734–2744.

[4] M. Doi, S.F. Edwards. *The theory of polymer dynamics*. Oxford University Press, New York, 1986.

[5] S.R. Durell, J.K. Labanowski, E.L. Gross. *Modeling of the electrostatic potential field of plastocyanin*. Arch. Biochem. Biophys., 277 (1990), 241–254.

[6] F. Fogolari, A. Brigo, H. Molinari. The Poisson-Boltzmann equation for biomolecular electrostatics: A tool for structural biology. *J. Mol. Recognit.*, 15 (2002), 377–392.

[7] R.R. Gabdoulline, R.C. Wade. Brownian Dynamics Simulation of Protein-Protein Encounter. *Methods*, 3 (1998), 329–341.

[8] A.B. Hope. Electron transfers amongst cytochrome f, plastocyanin and photosystem I: kinetics and mechanisms. *Biochim. Biophys. Acta*, 1456 (2000), 5–26.

[9] J. Janin. *Kinetics and thermodynamics of protein-protein interactions*. Protein-protein recognition. (Ed. C. Kleanthous). Oxford University Press, Oxford, 2000, 1–32.

[10] O.S. Knyazeva, I.B. Kovalenko, A.M. Abaturova, G.Y. Riznichenko, E.A. Grachev, A.B. Rubin. Multiparticle computer simulation of plastocyanin diffusion and interaction with cytochrome f in the electrostatic field of the thylakoid membrane. *Biophysics*, 55 (2010), No. 2, 221–227.

[11] I.B. Kovalenko, A.M. Abaturova, P.A. Gromov, D.M. Ustinin, E.A. Grachev, G.Y. Riznichenko, A.B. Rubin. Direct simulation of plastocyanin and cytochrome f interactions in solution. *Phys. Biol.*, 3 (2006), 121–129.

[12] I.B. Kovalenko, A.M. Abaturova, P.A. Gromov, D.M. Ustinin, G.Y. Riznichenko, E.A. Grachev, A.B. Rubin. Computer simulation of plastocyanin-cytochrome f complex formation in the thylakoid lumen. *Biophysics*, 53 (2008), No. 2, 140–146.

[13] I.B. Kovalenko, A.M. Abaturova, G.Y. Riznichenko, A.B. Rubin. A novel approach to computer simulation of protein-protein complex formation. *Dokl. Biochem. Biophys.*, 427 (2009), 215–217.

[14] I.B. Kovalenko, A.M. Abaturova, G.Y. Riznichenko, A.B. Rubin. Computer simulation of interaction of photosystem 1 with plastocyanin and ferredoxin. *BioSystems*, 103 (2010), 180–187.

[15] I.B. Kovalenko, A.N. Diakonova, A.M. Abaturova, G.Y. Riznichenko, A.B. Rubin. Direct computer simulation of ferredoxin and FNR complex formation in solution. *Phys. Biol.*, 7 (2010), No. 2, 026001.

[16] I.B. Kovalenko, O.S. Knyazeva, G.Y. Riznichenko, A.B. Rubin. Mechanisms of interaction of electron transport proteins in photosynthetic membranes of Cyanobacteria. *Doklady Biochemistry and Biophysics*, 440 (2011), 213–215.

[17] I.B. Kovalenko, O.S. Knyazeva, G.Y. Riznichenko, A.B. Rubin. Computer simulation of plastocyanin interaction with cytochrome f and photosystem I in cyanobacterium Phormidium laminosum. *Biophysics*, 59 (2014), 1–5.

[18] S. Niwa, H. Ishikawa, S. Nikai, T. Takabe. Electron transfer reactions between cytochrome f and plastocyanin from Brassica komatsuma. *J. Biochem.*, 88 (1980), 1177–1183.

[19] F. Panneton, P. L'Ecuyer. On the xorshift random number generators. *ACM T. Model. Comput. Sci.*, 15 (2005), No. 4, 346–361.

[20] D.C. Pearson, E.L. Gross. Brownian dynamics study of the interaction between plastocyanin and cytochrome f. *Biophys. J.*, 75 (1998), 2698–2711.

[21] F. Rienzo, R. Gabdoulline, M. Menziani, P. Benedetti, R. Wade. Electrostatic analysis and brownian dynamics simulation of the association of plastocyanin and cytochrome f. *Biophys. J.*, 81 (2001), 3090–3104.

[22] M. Ubbink, M. Ejdebeck, B.G. Karlsson, D.S. Bendall. The structure of the complex of plastocyanin and cytochrome f, determined by paramagnetic NMR and restrained rigid-body molecular dynamics. *Structure*, 6 (1998), 323–335.

[23] G.M. Ullmann, E.-W. Knapp. Electrostatic models for computing protonation and redox equilibria in proteins. *Eur. Biophys. J.*, 28 (1999), No. 7, 533–551.

[24] G.M. Ullmann, E.-W. Knapp, N.M. Kostic. Computational simulation and analysis of dynamic association between plastocyanin and cytochrome f. Consequences for the electron-transfer reaction. *J. Amer. Chem. Soc.*, 119 (1997), 42–52.

[25] U. Winter, T. Geyer. Coarse Grained Simulations of a Small Peptide: Effects of Finite Damping and Hydrodynamic Interactions. *J. Chem. Phys.*, 131 (2009), 104103.

In: Computational Materials ... ISBN: 978-1-63482-541-2
Editor: K. T. Kholmurodov, pp. 165-174

Chapter 15

Molecular–Dynamics Based Insights into the Problem of Graphite Melting

Nikita D. Orekhov[1,2,*] ***and Vladimir V. Stegailov***[1,2,3]
[1]Joint Institute for High Temperatures,
Russian Academy of Sciences,
Moscow, Russia
[2]Moscow Institute of Physics and Technology,
Dolgoprudny, Russia
[3]National Research University Higher School of Economics,
Moscow, Russia

Abstract

The experimental data on graphite melting temperature remain poorly determined despite the long history of investigations. The experimental results of several works cover the wide span from 3800 to 5000 K that is an essentially larger uncertainty than the errors of individual experiments. In this work we deploy the molecular dynamics (MD) method to study the kinetics of graphite melting, concerning the aspects of defect formation and single graphene layer melting. Our MD calculations show an unexpectedly weak kinetics of the melting front propagation in graphite that is several orders slower than that in metals. We demonstrate that at sufficiently high heating rates (higher than $10^5 - 10^6$ K/s) the temperatures 500-1000 K above the graphite melting temperature can be reached before the crystal decay. It allows us to explain long-standing problem of

*E-mail address: nikita.orekhov@gmail.com

the discrepancy in the experimental data making a hypothesis that there is a strong dependence between experimentally measured graphite melting temperatures and corresponding rates of heating.

Keywords: Molecular–Dynamics, Graphite Melting, Superheated Metastable, Solid Graphite

1. Introduction

Graphite is one of the most wide-spread carbon allotropes. Unique thermophysical properties of graphite result in its important role in science and engineering. However, the experimental data on graphite melting temperature still remain controversial despite the long history of investigation.

Investigations of graphite melting began with the experiment of F. P. Bundy (1963) on heating a sample by electric discharge [1]. Since that time, there have been several tens of studies directly concerning the position and behavior of the graphite melting curve; however, the question of graphite melting is still debated. The reason is the significant discrepancy of the results obtained. Experimentally obtained melting lines are located in a wide range of temperatures from 3700 to 5000 K [2, 1, 3, 4, 5, 6, 7, 8, 9, 10]. Detailed analysis of the main experimental researches from Pirani's [2] and Bundy's [1] papers to recent works was made in [11]. Experiments on graphite melting generally imply fast and ultrafast sample heating. The time scales of heating a sample and, accordingly, the processes occurring in it, are so small that they approach the limits that can be directly calculated by the molecular dynamics (MD) method. Thus, this method is a convenient tool for theoretical interpretation and supplementation of the experimental data, as well as for analyzing related phenomena. In this paper we report the results of MD simulation of graphite melting within the model of the empirical reactive interatomic potential, which describes the change in the type of hybridization of graphite atoms with a change in the structure of their local environment. The preliminary results have been presented in [12].

2. Model and Calculation Method

Calculation of the melting temperatures for various materials is of fundamental scientific interest. These calculations are conventionally carried out by finding

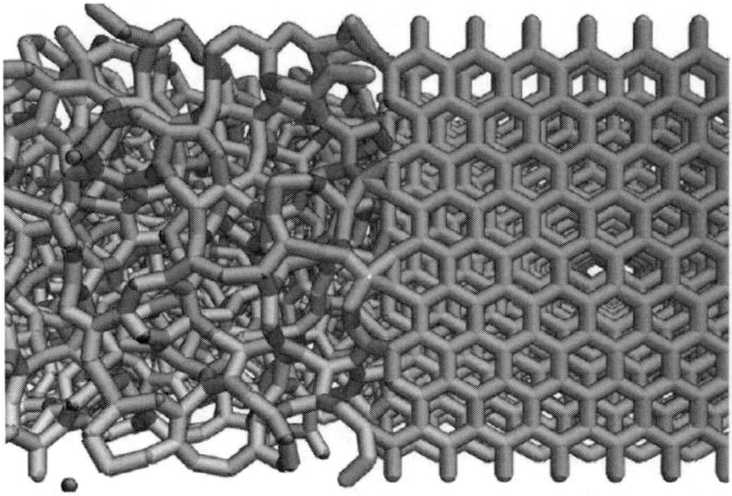

Figure 1. Fragment of the calculation cell with the melting front. The bonds between carbon atoms are shown (atoms are located in corresponding sites).

the free Gibbs energy for each phase with the subsequent search for equilibrium. An alternative method is the two-phase simulation method [13], which implies simulation of two coexisting phases; i.e., it suggests the simultaneous presence of both liquid and solid phases in the system under consideration. At a temperature close to the expected melting temperature, the solid partially passes to the liquid phase; the subsequent shift of the equilibrium between the phases allows one to draw conclusions about the melting temperature with respect to the initial material temperature. If the interface shifts to the disordered phase, the initial temperature exceeds the melting temperature. On the contrary, if the equilibrium shifts to the solid phase and crystallization begins, Modifications of this method were applied to various systems. To study the graphite melting kinetics, we carried out calculations for carbon atoms behaving according to the AIREBO potential [14]. The computational cell with periodic boundary conditions had sizes of 25, 25, and 120 along the OX, OY, and OZ axes, respectively (during simulation, the sizes were varied depending on pressure). About 10000 carbon atoms were used in the two-phase simulation. Half of the computational cell was occupied by a graphite crystal, while the other half was filled with disordered-phase atoms (Figure 1). To distinguish the ordered phase with respect to the disordered one, we used the static structural factor as an order parameter:

$$S = < cos(xk_x + yk_y + zk_z) >^2 + < sin(xk_x + yk_y + zk_z) >^2$$

where x, y, and z are the atomic coordinates and k_x, k_y, and k_z are the reciprocal lattice vectors corresponding to the graphite crystal structure. Averaging was performed over all atoms in the region where the structural factor was calculated. In the case of a disordered phase, the S value is close to zero, whereas for an ideal crystal it is unity. Using the structural factor, we monitored the position of the interface. We considered two variants of graphite crystal orientation: one with graphene layers normal (like in Figure 1) to the melting front and one with perpendicular. In practice, graphite is generally a polycrystalline material. Its structure, degree of isotropy, and density depend strongly on the preparation technique. The density may vary in the range of 1.6–2.2 g/cm^3, depending on the specific grade. Highly oriented pyrolytic graphite (HOPG), which is a quasi-monocrystal structure, composed of parallel graphene plates, is generally used in experiments on graphite melting. We specifically chose HOPG for calculations as a model for describing the solid phase. One calculation corresponded to 1.5–2 ns of time in most cases. The equations of atomic motion were integrated with a step of 0.1 fs. The system was brought to the desired parameters (temperature and pressure) using a NoseHoover thermostat and varying the cell sizes.

3. Results

The melting front velocities calculated for pressure P = 6 GPa are shown in Figure 2 in the range of temperatures T=3800–4700 K. There are many studies devoted to the calculation of the dependence of the interface velocity on the degree of superheating (supercooling). Note that the melting kinetics for graphite is slower than that for metals by about two orders of magnitude [15]. The front velocity is only about 1 m/s in the range of 4400–4700 K (corresponding superheating more than 600–900 K), which is close in order of magnitude to the corresponding value for silicon [16]. To verify whether such slow rate of melting is some specific effect of AIREBO potential or not, we have performed the same calculations with another accurate bond-order carbon potential - LCBOP[17]. According to LCBOP–I model graphite melting temperature lays in the range 3750–4000 K [18]. Later modification of this potential - LCBOP–II gave slightly higher value of graphite melting temperature T_m=4250 K [19].

Nevertheless our calculations with LCBOP–I show the same slow melting kinetics and even higher levels of possible sample overheating: crystal maintained its structure at temperatures $T > 5000$. The relatively low velocity of the melting front may affect, in particular, the results of the experiments on heating graphite by a short (shorter than one millisecond) current pulse, where the melting onset is determined from the change in the macroscopic sample resistance.

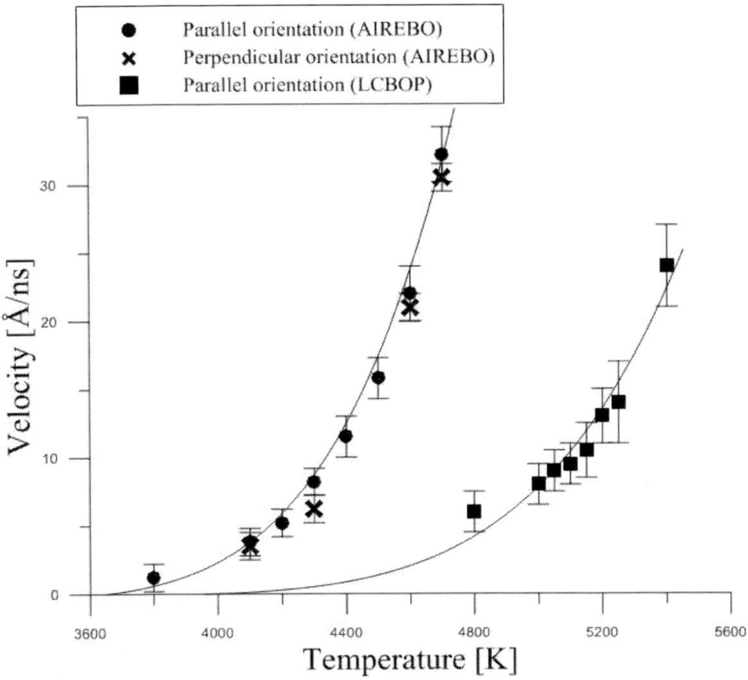

Figure 2. Temperature dependence of the melting front velocity for AIREBO and LCBOP potentials.

In the temperature range of 3750-3950 K, the front velocity decreases values close to zero. It is difficult to determine exactly the instant of turning of the front velocity to zero because of the slow rate of the process and, accordingly, large calculation time. A further slow decrease in temperature revealed no tendencies to crystallization onset at times on the order of a few nanoseconds.

Graphite melting at high levels of superheating is primarily initiated by the formation of covalent bonds between atoms located in different layers. Figure 3 shows the dynamics of the development of such defects according to the homogeneous nucleation mechanism at strong overheating in an initially ideal crystal, which leads to lattice melting. Similar processes occur near the interface (i.e., they facilitate heterogeneous melting).

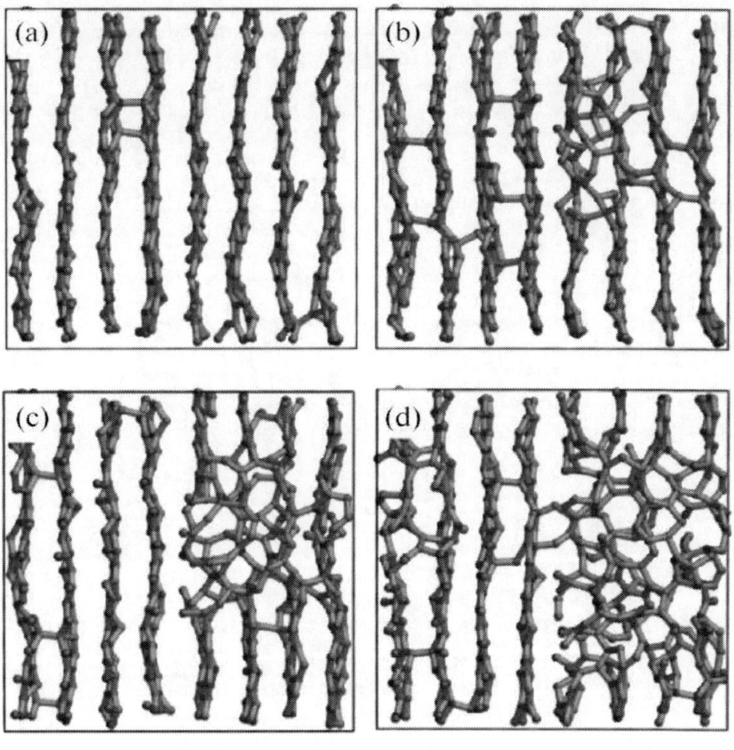

Figure 3. Dynamics of the formation of a group of interlayer defects initiating melting in the bulk of an ideal graphite crystal at T = 4700 K and P = 12 GPa at (a) 1.6, (b) 2.1, (c) 2.3, and (d) 2.5 ns after the beginning of calculation.

Liquid phase formation at temperatures below 4300–4400 requires the presence of larger defects (in particular, grain boundaries, characteristic of polycrystalline graphite, or liquid-solid interfaces) in the crystal. In the absence of considerable defects and homogeneities, the crystal can withstand a significant overheating on finite time scales. For example, at a pressure of 12 GPa, sample melting in the cell containing initially no liquid phase was observed at times

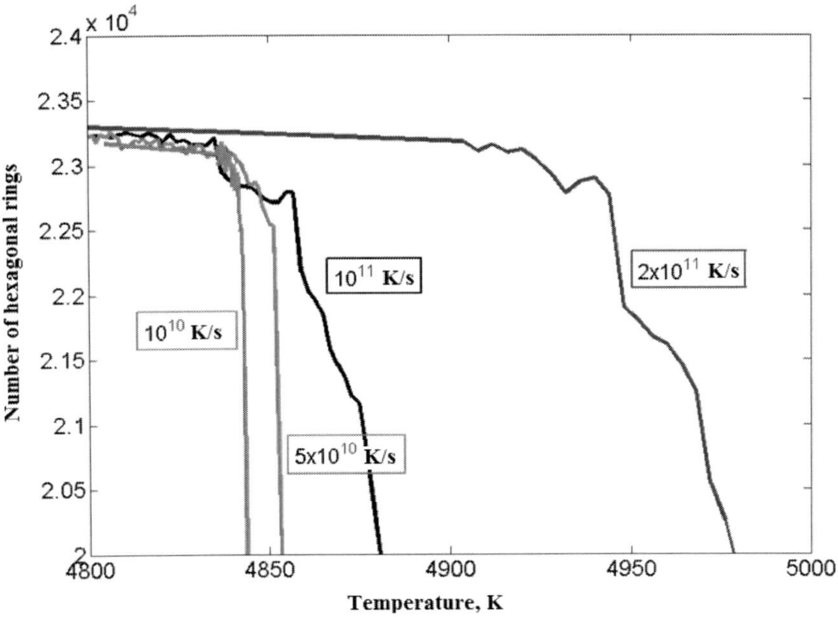

Figure 4. Dynamics of graphene thermal decay. Heating rates are indicated near the curves.

below 10 ns only at temperatures exceeding 4650 K (overheating by more than 850 K). At temperatures below 4600 K, the crystal structure remained stable up to the maximum calculation time (10 ns). In the case of pressures of 2 GPa, melting in the initially ideal lattice was not observed even at T = 4800 K (overheating by about 1000 K), which is close to the graphene melting temperature. Thus, the close-to-ideal graphite lattice may retain its structure in the metastable region at a significant superheating on nanosecond time scales. Taking into account the suggestions about the dependence of the melting temperature on the sample heating time, which were reported in [9], the results obtained may clarify some factors leading to discrepancies in the experimental data.

At temperatures above 4800–4850 K, the mechanism of graphite melting changes. In this range, some portions of graphene layers begin to fracture, irrespective of the initial position of the phase interface (homogeneous nucleation [20]). The aspects of graphene melting were considered in detail in [21]; it was shown (using the Monte Carlo method and the LCBOPII potential) that a graphene sheet successfully retains its structure upon heating to 4900 K. For the AIREBO potential under the same conditions, we obtained that the graphene melting temperature is also in the range of 4800–4900 K.

Graphene melting is driven by the formtion of the clusters of Stone-Wales defects. Dynamics of graphene structure decay is shown in Figure 4. During the constant rate heating at temperatures above 4800 K the portion of hexagonal rings begin to decrease rapidly. Melting begins at the points with highest Stone-Wales defect concentration and results in formation of long linear carbon chains. The similar picture was observed for LCBOP in [21].

We consider the temperature of graphene melting T=4800–4900 K to be the upper bound of the value of metastable graphite zone, since graphene represents the two-dimensional crystal with no interlayer defects. Note that this value of graphene melting temperature exceeds significantly the graphite melting temperatures obtained for LCBOP in [18, 19] and our results for AIREBO; however, it coincides with a high accuracy with the upper range of experimentally found graphite melting temperatures. Having disregarded the van der Waals interlayer interaction, whose energy is low in comparison with the covalent interaction, a single graphene layer can be considered as an element of graphite crystal, the melting of which is not related to the formation of interlayer defects.

4. Conclusion

Due to the difference between melting temperatures obtained with AIREBO and LCBOP–II in [19] we can make only approximate suggestions about the value of the graphite melting temperature. Nevertheless in our calculations we observed a more interesting fact: this two independent models of carbon potentials show similar surprisingly slow kinetics of graphite melting. Also it was noted that an ideal graphite structure can withstand large overheatings (up to 500–1000 K) during times longer than 10 ns, especially at relatively low pressures $P < 2 - 3$ GPa (a range where the concentration of interlayer defects is low). The results presented can be considered as an essential argument in favor of the hypothesis that at $T > 10^5 - 10^6$ K/s the de facto detected in pulse heating experiments graphite melting temperatures T_m^d should be considered as the temperatures when the decay of the superheated metastable solid graphite takes place. And therefore these temperatures are not strictly speaking the graphite melting temperatures in the thermodynamic sense T_m.

References

[1] F.P. Bundy. Melting of Graphite at Very High Pressure. *The Journal of Chemical Physics*, 38(3):618, 1963.

[2] M S von Pirani. Elektrothermie: die elektrische Erzeugung und technische Anwendung hoher Temperaturen. *J. Springer*, pages 1–557, 1930.

[3] N.S. Fateeva. Concerning the Melting Point of Graphite up to 90 kbar. *ZhETF Pisma v Redaktsiiu*, 13:157–159, 1971.

[4] A.V. Kirillin, M.D. Kovalenko, M.A. Sheindlin, and V.S. Zhivopistsev. Stationar laser heating in measuring the vapor pressure of carbon at 5000-7000 K. *Templofiz Vys Temp*, 23(4)(21):699–706, June 1985.

[5] G. Pottlacher, R.S. Hixson, S. Melnitzky, E. Kaschnitz, M.A. Winkler, and H. Jäger. Thermophysical properties of poco axf-5q graphite up to melting. *Thermochimica acta*, 218:183–193, 1993.

[6] C. Ronchi, R. Beukers, H. Heinz, J. P. Hiernaut, and R. Selfslag. Graphite melting under laser pulse heating. *International Journal of Thermophysics*, 13(1):107–129, January 1992.

[7] M. Togaya. Pressure Dependences of the Melting Temperature of Graphite and the Electrical Resistivity of Liquid Carbon. *Physical Review Letters*, 79(13):2474–2477, September 1997.

[8] M. Musella, C. Ronchi, M. Brykin, and M. Sheindlin. The molten state of graphite: An experimental study. *Journal of Applied Physics*, 84(5):2530, 1998.

[9] E.I. Asinovskii, A.V. Kirillin, and A.V. Kostanovskii. Experimental investigation of the thermal properties of carbon at high temperatures and moderate pressures. *Physics-Usp*, 45(8):869–882, August 2002.

[10] A.I. Savvatimskiy. Melting point of graphite and liquid carbon. *Phys-Usp*, 173(12)(21):1295–1303, June 2003.

[11] A.I. Savvatimskiy. Measurements of the melting point of graphite and the properties of liquid carbon (a review for 1963?2003). *Carbon*, 43(6):1115–1142, May 2005.

[12] N.D. Orekhov and V.V. Stegailov. Molecular dynamics simulation of graphite melting. *High. Temp.*, 52(2):220–228, 2014.

[13] J. Morris, C. Wang, and K. Ho. Relationship between structure and conductivity in liquid carbon. *Physical Review B*, 52(6):4138–4145, August 1995.

[14] S.J. Stuart, A.B. Tutein, and J.A. Harrison. A reactive potential for hydrocarbons with intermolecular interactions. *The Journal of Chemical Physics*, 112(14):6472, 2000.

[15] A.Y. Kuksin, I.V. Morozov, G.E. Norman, V.V. Stegailov, and I.A. Valuev. Standards for molecular dynamics modelling and simulation of relaxation. *Molecular Simulation*, 31(14-15):1005–1017, December 2005.

[16] D. Buta, M. Asta, and J. Hoyt. Kinetic coefficient of steps at the Si (111) crystal-melt interface from molecular dynamics simulations. *The Journal of chemical physics*, 127(7):074703, 2007.

[17] J. Los and A. Fasolino. Intrinsic long-range bond-order potential for carbon: Performance in monte carlo simulations of graphitization. *Physical Review B*, 68(2):024107, 2003.

[18] L. Ghiringhelli, J. Los, E. Meijer, A. Fasolino, and D. Frenkel. Modeling the Phase Diagram of Carbon. *Physical Review Letters*, 94(14):145701, April 2005.

[19] F. Colonna, J. Los, a. Fasolino, and E. Meijer. Properties of graphite at melting from multilayer thermodynamic integration. *Physical Review B*, 80(13):1–8, October 2009.

[20] V.V. Stegailov. Homogeneous and heterogeneous mechanisms of superheated solid melting and decay. *Computer physics communications*, 169(1):247–250, 2005.

[21] K.V. Zakharchenko, A. Fasolino, J.H. Los, and M.I. Katsnelson. Melting of graphene: from two to one dimension. *Journal of Physics: Condensed Matter*, 23(20):202202, 2011.

In: Computational Materials ... ISBN: 978-1-63482-541-2
Editor: K. T. Kholmurodov, pp. 175-185

Chapter 16

PHASE DIAGRAMS OF GAS HYDRATES: MOLECULAR SIMULATION STUDIES

***Grigory S. Smirnov*[1,2,*] *and Vladimir V. Stegailov*[1,2,3]**
[1]Joint Institute for High Temperatures,
Russian Academy of Sciences,
Moscow, Russia
[2]Moscow Institute of Physics and Technology,
Dolgoprudny, Russia
[3]National Research University Higher School of Economics,
Moscow, Russia

Abstract

In this work, we perform coexistence simulations of methane hydrates for pressures up to 5000 bar for different water models. We calculate the kinetic stability boundary of the superheated metastable sI structure and analyze the effects of the heating rate, system size and cage occupancy. We also report molecular dynamics simulation of several possible structures for the new hydrogen hydrate clathrate. We show the strength of molecular simulation as a supplement tool for the analysis of experimental data

Keywords: Coexistence Simulations, Methane Hydrates, High Pressures, Water Models, Molecular Dynamics Simulations

[*]E-mail address: grsmirnov@gmail.com

1. Introduction

Gas hydrates are crystalline water-based inclusion compounds physically resembling ice [1]. The discovery of hydrogen hydrates (HH) [2, 3] attracted significant attention to the H_2+H_2O phase diagram and clathrate structures. Along with the fundamental interest and significance for geophysics of icy moons and outer planets hydrogen hydrates provide a way to perspective hydrogen storage technologies. The wide-spread methane hydrate forms relatively well-studied sI, sII and sH clathrate structures. The H_2+H_2O system is known to form sII structure and two "filled-ice" structures C_1 and C_2.

The importance of methane and hydrogen hydrates requires the accurate knowledge of their thermodynamic and kinetic properties, mechanisms of formation and decay. Molecular simulation is a method of choice for such theoretical studies since it can explicitly capture the structure of gas hydrates and their constituents.

2. Methane Hydrates

2.1. Model

The state-of-art classical potentials allow quite accurate overall description of the water phase diagram in the solid phase [4]. Important features of water can be captured even by simplified potential models.We use the TIP4P/Ice model that gives a very good description of the ice phases coexistence lines. For comparison we consider TIP4P/2005 and a well-known SPC/E model. SPC/E is a simple 3-site model with charges located on H and O sites. TIP4P models are the 4 site models with a negative massless charge located near the oxygen atom and positive charges located on H atoms. We use a simple Lennard-Jones (LJ) model for methane proposed by Guillot and Guissani [5] and Paschek [6].

The cross interaction between methane and water is described by the Lorentz-Berthelot rules

$$\varepsilon_{\mathrm{CH_4-H_2O}} = \chi\sqrt{\varepsilon_{\mathrm{CH_4-CH_4}}\varepsilon_{\mathrm{H_2O-H_2O}}},$$

$$\sigma_{\mathrm{CH_4-H_2O}} = \frac{\sigma_{\mathrm{CH_4-CH_4}} + \sigma_{\mathrm{H_2O-H_2O}}}{2},$$

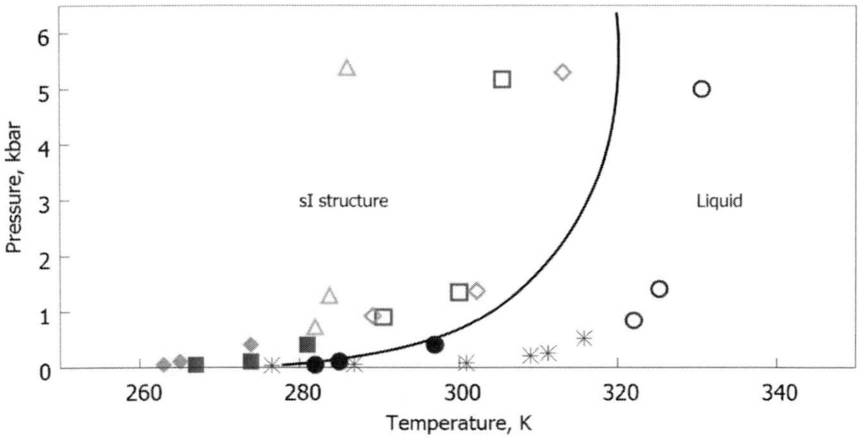

Figure 1. Methane-water phase diagram. The solid line is the experimental [9] three-phase equilibrium curve of methane hydrate. The snowflakes show the result of Jensen et al. [10] for TIP4P/Ice model. The filled blue, green and red symbols show the three-phase coexistence points of Conde and Vega [11] and open symbols show our results: yellow triangles – SPC/E, green diamonds and red squares – TIP4P/2005 with $\chi = 1.07$ and 1.00 respectively, blue circles – TIP4P/Ice. Our symbols correspond to $5 \times 5 \times 10$ unit cell systems.

where $\chi = 1$ or 1.07. The latter value describes the chemical potential of methane in TIP4P/2005 water better [7]. All MD calculations are performed using LAMMPS [8].

2.2. Equilibrium Melting

First we determine coexistence equilibrium temperatures and pressures. Conde and Vega in their work [11] performed similar calculations using long NPT MD trajectories (up to 1 μs). At several fixed temperatures they waited for complete crystallization or complete melting of the initial three-phase system. We follow another approach looking directly for the phase coexistence conditions [12, 13]. At the preliminary stage in our calculations one half of the system is melted while another half is frozen. Then a short NPT simulation is performed to drive the system to the desired temperature and pressure. Finally the system evolve for several nanoseconds in the NVE MD run. Depending on the initial kinetic energy of the system one can observe either melting or growth of the sI phase. Reaching the equilibrium during crystallization requires longer simulation times, especially when methane molecules form a bubble in water. In our calculations we consider only those cases when hydrate melts and equilibrium establishes much faster, during a few nanoseconds. When the volume of the sI phase stops changing we assume that the temperature and pressure in the system correspond to the coexistence conditions.

The typical simulation box in the previous studies of gas hydrates contains $2\times2\times2$ or $2\times2\times4$ unit cells. In our simulation we use larger system sizes to decrease the possible influence of size effects. Our results for different water models (TIP4P/Ice, TIP4P/2005 and SPC/E) are shown in Fig. 1.

According to Conde and Vega [11], the TIP4P/Ice model provides the best agreement with the experimental data. Jensen et al. [10] determined the sI melting line by free energy calculations via Monte Carlo method for TIP4P/Ice model and the agreement of their results is worse than it was found by Conde and Vega (although the LJ potentials for methane were slightly different). Our MD results are in agreement with the data of Jensen et al. [10]. This is a strange fact because our results for TIP4P/2005 models are in a fairly good agreement with Conde and Vega [11]. We attribute this discrepancy to the larger interface area of our model (5×5 unit cells compared to 2×2 in Ref. [11]). Presumably, smaller interface cross-sections can result in larger statistical uncertainty and biased coexisting pressure and temperature values. Our results show that the TIP4P/2005 model gives the better agreement with the experimental coexistence line than the TIP4P/Ice model in the entire pressure range considered. The coexistence temperature values are systematically 10-20 K lower for the former model but the qualitative curve shape reproduces the experimental data quite well.

3. Hydrogen Hydrates

In this part we discuss new variants of the clathrate structures. Recently a new clathrate phase of water-hydrogen system was reported by Efimchenko et al. [14] at about 253 K and 5 kbar and by Strobel et al. [15] at about 170 K and 7 kbar. X-ray diffraction patterns do not allow to determine its structure unambiguously. In this work we also discuss the experimental puzzle of the new HH structure using atomistic computational models [16].

The novel C_0 structure is not completely resolved. As a starting point Efimchenko et al. proposed [14] that this structure has 7.5 water and 1.5 hydrogen

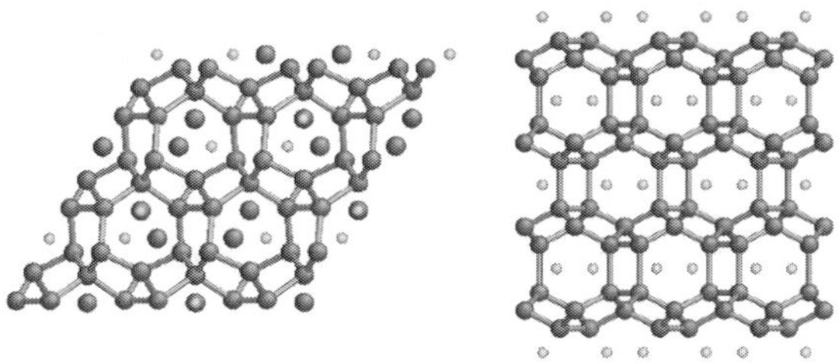

Figure 2. C_0-II and sT′ clathrate structures. Sticks show the cages formed by the hydrogen bonds between water molecules that are located in the nodes. H_2 molecules are represented by grey spheres which radius equals to the H_2 LJ radius. Cages in C_0-II structure actually form the channels.

molecules per unit cell. It should be noted that half-occupied positions previously were not observed either for hydrates or ices. A possible explanation is that Efimchenko et al. observed nitrogen instead of oxygen at the $3a_2$ position (M. Kuzovnikov, unpublished results) because N_2 molecules may diffuse into the sample during storage in liquid nitrogen. Therefore, a “true” C_0 structure can be comprised of 6 water and 3 hydrogen molecules per unit cell providing hypothetically a high level of molecular hydrogen storage with 5.6 wt % at maximum. To distinguish these two variants of C_0 structure we denote the original variant of Efimchenko et al. with half-occupied oxygen positions as C_0-I and the variant with hydrogen molecules at $3a_2$ sites as C_0-II (see Figure 2).

The α-quartz structure has 9 water molecules per unit cell. Using the lattice constants from Strobel et al. [15], we find that the average distance between oxygen atoms is about 2 Å. That is why pressure in this structure should be unrealistically high. The increase of lattice constants a and c would lead to disagreement with the experimentally measured X-ray pattern. sT′ structure contains $4^2 5^8 6^4$ cavities with 12 water and 4 hydrogen molecules per unit cell (see Figure 2).

3.1. Model

As discussed earlier, building of atomistic models for such supramolecular systems as gas hydrates is essentially based on the interatomic interaction models.

Frankcombe and Kroes [17] studied decomposition of the hydrogen-containing sII clathrate under constant temperature conditions. They compared different water models and found that TIP5P model is in the best agreement with experiment. In this work we use TIP4P/2005 [18] and TIP4P/Ice [19] water models that give the best description of the phase diagram of water in the solid state [20, 21].

Several methods for adding molecular hydrogen into the forcefield were proposed. One of the widely used forcefields for H_2 was developed by Alavi et al. [22]. In this work we use two different models for hydrogen: one is a united-atom Lennard-Jones (UA LJ) model [23, 24] and another is a three-site potential by Alavi et al [22]. The cross-interaction between hydrogen and water is described by the Lorentz-Berthelot rules.

3.2. Results

Phase stability studies generally proceed from the principle that the stable structure has the lowest free energy at a given set of conditions.

However, the accurate calculation of the free energy for multicomponent systems is a very difficult and challenging task [25]. Therefore in order to make an educated guess about the stability of the structures, we calculate the potential energy of the corresponding empty lattices formed of water molecules without hydrogen molecules at zero temperature. Energy per water molecule for each structure as a function of volume is shown in Figure 3. It can be seen that potential energy for C_0-II and sT$'$ structures are close, but sT$'$ is somewhat denser packed. C_0-I and α-quartz structures have higher energy. This analysis clearly disqualifies the α-quartz HH structure.

The next tool to study the phase stability is direct MD simulations of the HH structures at different pressures and temperatures.

The simulation box is orthogonal and contained 1350 H_2O and 270 H_2 molecules for C_0 structure, 1350 H_2O and 300 H_2 molecules for α-quartz, 900 H_2O and 300 H_2 molecules for sT$'$. Simulation procedure is the following: each run consists of a short 100 ps NPT simulation followed by a long 10-50 ns NVE simulation. During the NPT run the system achieves desired temperature and pressure. Then the pressure is determined by the fixed size of the MD simulation cell in the NVE ensemble that is superior to NPT ensemble for studying metastable phase stability. Since the system has been equilibrated the pressure and temperature remain constant during the NVE MD run until the initial structure decays (in the cases it does).

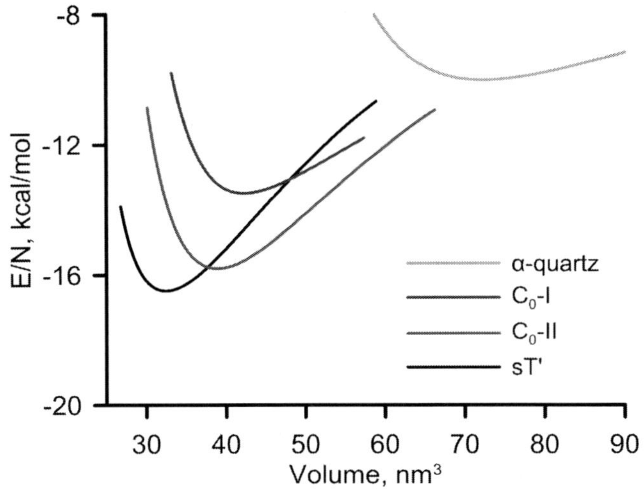

Figure 3. Potential energy per water molecule as a function of volume for the corresponding empty structures formed of water molecules without hydrogen molecules at zero temperature (TIP4P/Ice model).

Decomposition processes in CH_4 hydrates were studied in our work [26]. The average lifetime of metastable states depends on different factors: the degree of metastability, the number of molecules in the model, the choice of interaction potential, the H_2/H_2O ratio, etc. It is well known that the limited size of the MD simulation box decreases the probability of new phase nucleation. Therefore in MD simulations truly thermodynamically stable states can be hardly distinguishable from metastable states. However, even the evidence of the limited stability of the given phase (e.g., for 50 ns of time) is enough to claim that this structure can possess its region of stability.

We find that C_0-I (with $3a_2$ water molecules) fully decomposes during the first 10-20 ps of the NPT runs for all interatomic interaction models under consideration. Therefore, only sT′ and C_0-II structures (with hydrogen instead of $3a_2$ water molecules) can be the candidates for the further study.

The sT′ structure shows no sign of decomposing after 50 ns at (140 K; 2 kbar), (140; 10), (220; 6) and (300; 2). This fact holds for all the forcefield combinations under consideration. At (300 K; 10 kbar) the sT′ behaviour depends on interatomic interaction model: the sT′ phase is only stable for TIP4P/Ice and 3-site H_2 models, fully decomposes at the end of NPT runs for TIP4P/2005 model and partially decomposes at the end of NVE run for TIP4P/Ice and UA LJ models.

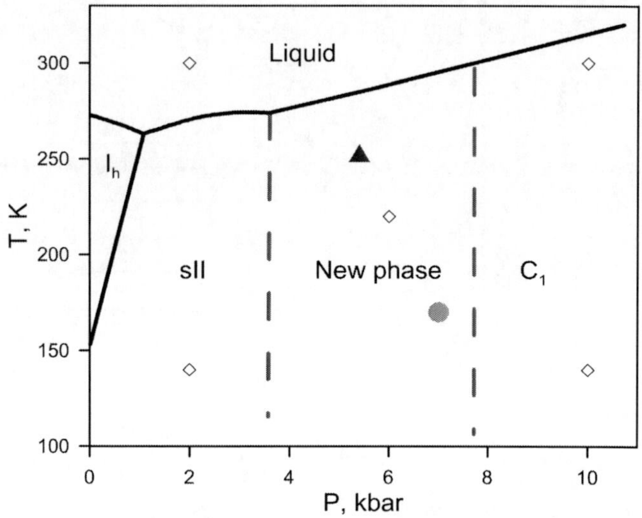

Figure 4. The hydrogen–water phase diagram. Summary of the availible experimental data are shown as solid lines. Dashed lines show the possible location of the new phase. A blue triangle and the green circle show the synthesis conditions in the experiments of Efimchenko et al. [14] and Strobel et al. [15]. Open diamonds are the phase points considered in this work.

The C_0-II structure shows somewhat higher stability in comparison with the sT′ structure. It is unstable only at (300 K; 2 kbar) for TIP4P/2005 model. In 4 other (T,P) phase points and for other forcefield combinations the sT′ structure does not decay during all 50 ns MD runs. For the combination of TIP4P/Ice and UA LJ models the MD runs are extended to 50 ns.

So we see that both C_0-II and sT′ structures show stability at the pressures and temperatures that correspond to the new HH phase. Presumably in experiments a mix of sT′ and C_0-II phases could be observed.

References

[1] E. Dendy Sloan and Carolyn Koh. *Clathrate Hydrates of Natural Gases, Third Edition.* CRC Press, Boca Raton, FL, 2007.

[2] Willem Vos, Larry Finger, Russell Hemley, and Ho-Kwang Mao. Novel H_2-H_2O Clathrates at High Pressures. *Phys. Rev. Lett.*, 71(19):3150–3153, November 1993.

[3] Yu. A. Dyadin, É. G. Larionov, E. Ya. Aladko, A. Yu. Manakov, F. V. Zhurko, T. V. Mikina, V. Yu. Komarov, and E. V. Grachev. Clathrate Formation in Water-Noble Gas (Hydrogen) Systems at High Pressures. *J. Struct. Chem.*, 40(5):790–795, September 1999.

[4] C Vega, E Sanz, J L F Abascal, and E G Noya. Determination of phase diagrams via computer simulation: methodology and applications to water, electrolytes and proteins. *Journal of Physics: Condensed Matter*, 20(15):153101, April 2008.

[5] B. Guillot and Y. Guissani. A computer simulation study of the temperature dependence of the hydrophobic hydration. *J. Chem. Phys.*, 99(10):8075–8094, 1993.

[6] Dietmar Paschek. Temperature dependence of the hydrophobic hydration and interaction of simple solutes: An examination of five popular water models. *J. Chem. Phys.*, 120(14):6674–6690, 2004.

[7] H. Docherty, A. Galindo, C. Vega, and E. Sanz. A potential model for methane in water describing correctly the solubility of the gas and the properties of the methane hydrate. *J. Chem. Phys.*, 125(7):074510, 2006.

[8] Steve Plimpton. Fast parallel algorithms for short-range molecular dynamics. *J. Comput. Phys.*, 117(1):1–19, 1995.

[9] Y. A. Dyadin and E. Y. Aladko. Decomposition of the methane hydrate up to 10 kbar. In J.P. Monfort, editor, *Proceedings of the Second International Conference on Natural Gas Hydrates*, 1996.

[10] Lars Jensen, Kaj Thomsen, Nicolas von Solms, Scott Wierzchowski, Matthew R. Walsh, Carolyn A. Koh, E. Dendy Sloan, David T. Wu,

and Amadeu K. Sum. Calculation of liquid water-hydrate-methane vapor phase equilibria from molecular simulations. *J. Phys. Chem. B*, 114(17):5775–5782, 2010.

[11] M. M. Conde and C. Vega. Determining the three-phase coexistence line in methane hydrates using computer simulations. *J. Chem. Phys.*, 133(6):064507, 2010.

[12] J. R. Morris, C. Z. Wang, K. M. Ho, and C. T. Chan. Melting line of aluminum from simulations of coexisting phases. *Phys. Rev. B*, 49(5):3109–3115, Feb 1994.

[13] Sergey V. Starikov and Vladimir V. Stegailov. Atomistic simulation of the premelting of iron and aluminum: Implications for high-pressure melting-curve measurements. *Phys. Rev. B*, 80(22):220104, Dec 2009.

[14] V.S. Efimchenko, M.A. Kuzovnikov, V.K. Fedotov, M.K. Sakharov, S.V. Simonov, and M. Tkacz. New Phase in the Water-Hydrogen System. *J. Alloys Compd.*, 509:S860–S863, September 2011.

[15] Timothy A. Strobel, Maddury Somayazulu, and Russell J. Hemley. Phase Behavior of H_2 + H_2O at High Pressures and Low Temperatures. *J. Phys. Chem. C*, 115(11):4898–4903, March 2011.

[16] Grigory S Smirnov and Vladimir V Stegailov. Toward Determination of the New Hydrogen Hydrate Clathrate Structures. *The Journal of Physical Chemistry Letters*, 4(21):3560–3564, November 2013.

[17] T.J. Frankcombe and G.-J. Kroes. Molecular Dynamics Simulations of Type-sII Hydrogen Clathrate Hydrate Close to Equilibrium Conditions. *J. Phys. Chem. C*, 111(35):13044–13052, September 2007.

[18] J. L. F. Abascal and C. Vega. A General Purpose Model for the Condensed Phases of Water: TIP4P/2005. *J. Chem. Phys.*, 123(23):234505, December 2005.

[19] J. L. F. Abascal, E. Sanz, R García Fernández, and C. Vega. A Potential Model for the Study of Ices and Amorphous Water: TIP4P/Ice. *J. Chem. Phys.*, 122(23):234511, 2005.

[20] C Vega, E Sanz, J L F Abascal, and E G Noya. Determination of Phase Diagrams via Computer Simulation: Methodology and Applications to Water, Electrolytes and Proteins. *J. Phys.: Condens. Matter*, 20(15):153101, April 2008.

[21] M M Conde and C Vega. Note: A Simple Correlation to Locate the Three Phase Coexistence Line in Methane-Hydrate Simulations. *J. Chem. Phys.*, 138(5):056101, 2013.

[22] Saman Alavi, J A Ripmeester, and D D Klug. Molecular-Dynamics Study of Structure II Hydrogen Clathrates. *J. Chem. Phys.*, 123(2):24507, July 2005.

[23] D. Sabo, S. B. Rempe, J. A. Greathouse, and M. G. Martin. Molecular Studies of the Structural Properties of Hydrogen Gas in Bulk Water. *Mol. Simul.*, 32(3-4):269–278, March 2006.

[24] S Mondal, S Ghosh, and P K Chattaraj. A Molecular Dynamics Study on sI Hydrogen Hydrate. *J. Mol. Model.*, pages 20–24, October 2012.

[25] Lars Jensen, Kaj Thomsen, Nicolas von Solms, Scott Wierzchowski, Matthew R. Walsh, Carolyn A. Koh, E. Dendy Sloan, David T. Wu, and Amadeu K. Sum. Calculation of liquid water-hydrate-methane vapor phase equilibria from molecular simulations. *J. Phys. Chem. B*, 114(17):5775–5782, 2010.

[26] Grigory S. Smirnov and Vladimir V. Stegailov. Melting and Superheating of sI Methane Hydrate: Molecular Dynamics Study. *J. Chem. Phys.*, 136(4):044523, 2012.

INDEX

A

B

C

D

F

G

J

L

M

N

O

P

Q

R

S

T

V

X